Geographical Information Systems (GIS) in Library and Information Services

About the Author

Dattatray Narayan Phadke (*b.* 1949). Graduated in Chemistry (1972). He entered into the field of Library and Information Science after doing M.Sc.(Physical Chemistry) from University of Pune (1973). Completed M.Lib. from University of Mumbai (1986) and subsequently Ph.D. in Library and Information Science from University of Pune with the topic "Use of GIS in Library" (1999).

Dr. Phadke after initially working for a few years (VRDE, DRDO), Ahmednagar moved on to Central Library IIT Bombay (1979). Since than he has been serving at IIT Bombay at different levels associated with various activities of the Library. Currently he is one of the Senior Assistant Librarians. He also worked on MODROB / institute sponsored project on Providing Comprehensive Electronic Security to Valuable Library Collection.

Dr. Phadke has contributed a number of articles in journals, seminars and conferences. He has a book edited entitled "In Search of Excellence—Librarianship" published in the honour of his guide Dr. S.R. Ganpule by Multi-Tech Pub. Co., Mumbai (1999).

As an expert in Library and Information Science he is associated with M.Lib.Sc. Programme of YCMOU, Nasik, where he has contributed a number of course books and workbooks on SOUL software of INFLIBNET, Ahmedabad. He is recognized as Ph.D. Guide by YCMOU, Nasik and a few other Universities.

Geographical Information Systems (GIS) in Library and Information Services

D.N. PHADKE

Asst. Librarian (S.G.)

Central Library, Indian Institute of Technology, Bombay

CONCEPT PUBLISHING COMPANY, NEW DELHI-110059

Cataloging in Publication Data—DK
Courtesy: D.K. Agencies (P) Ltd. <docinfo@dkagencies.com>

Phadke, D. N. (Dattatray Narayan), 1949-
Geographical information systems (GIS) in library and information services / D.N. Phadke.
p. cm.
Includes bibliographical references.
Includes index.
ISBN 818069271X

1. Libraries—Geographic information systems.
I. Title.

026.910 285 22

ISBN : 81-8069-271-X

First Published 2006

Published and Printed by
Ashok Kumar Mittal
Concept Publishing Company
A/15-16, Commercial Block, Mohan Garden,
New Delhi-110059 (India)

Phones : 25351460, 25351794
Fax : 091-11-25357103
Email : publishing@conceptpub.com

DEDICATED
TO
MY BELOVED FATHER NANA,
MOTHER AND MY FAMILY
KUMUD, DEPALI AND SOURABH

PREFACE

Geographical Information Systems (GIS) has traditionally been used in areas like land-use, urban planning, environmental and demographic analysis, natural resources assessment etc. Now GIS finds its applications in non-traditional areas like business, banking, health and other utility services also. Researchers are trying to explore GIS applications in many other areas concerned with human life. GIS is useful in all these applications because of its spatial data handling and efficient data management characteristics. Its special feature of linking map data/spatial data with attribute data facilities analysis, modelling and research by using geographic dimension. This makes it possible to use GIS in almost all fields where spatial or map data and attribute data come into the picture. On the similar lines it was thought that GIS might find use in the library also.

The present study attempts to apply Geographical Information Systems (GISs), specially its spatial data handling characteristics, in the library environment. The application is mainly with a view to provide a tool for finding physical location of library material by handling huge locational data in a large library. The study, in the process, probes into other uses of GIS in library. It then discusses the prospects of such use.

In the first phase, the study discusses the core or the important use of GIS in library viz, "Use of GIS in Mapping Library Collection" at a greater length. Subsequently, the attempt has also been made to demonstrate the concept, through the development of a prototype of GIS based library system. Prototype results, presented with examples, clearly explain how one can map the collection and find out the actual location of library material in a digitised map of the library. It is pointed out that this can also be an additional feature to OPAC.

With the support of use of GIS referred to above, the study further discusses two other important applications of GIS viz. (a) Use of GIS in Mapping the network of Libraries and (b) Use of GIS in providing library services using spatially referenced databases.

In the third phase the study investigates all the other applications of GIS in the library. They include the use of GIS : (i) in mapping geographic facets against the subject content of books as assigned while classifying them by colon and other classification schemes, (ii) in mapping library book suppliers and distributors, (iii) in user study, and (iv) in user educational programmes.

In the course of present study many new ideas and concepts have emerged and as such study claims the following findings and original contributions :

- The study has proved that Geographical Information Systems (GIS) can be used in handling the geographical location data of library material which is tracked separately through the concept of "Mapping of Library Collection" and providing location finding tool in OPAC. This new concept has been demonstrated through development of phototype.
- The study has discussed new concept of automating shelving process of the library by introducing idea of "shelving model".
- While discussing use of GIS in Mapping the network of libraries the study has presented new concept of thematic mapping of library resources and use of internet for the purpose.
- Out of other uses of GIS in library the study contributes new concept of mapping geographic facets against subject contents by exploiting geographic treatment given by the Colon and other classification schemes.
- While identifying all the other probable uses of GIS in library the study has attempted to bring a new geographic dimension in all these areas which was not conceptualised so far. The study has introduced new terms like "Virtual Browsing" and "Virtual Library Network".

The study has been organised in six chapters. They are summarised below :

The first chapter discusses issues relating to the topic such as the concept of GIS, definitions, working of GIS technology, special characteristics of GIS and its various applications in different areas. It then discusses the relevance of GIS to the library in the light of problems related to areas where it could be applied. The main objective of the study is to explore the various ways in which GIS can be used in the library. Specific aims, objectives and methodology applied in the work are also discussed in this introductory chapter.

The review of literature related to the research topic is presented in the second chapter. Literature search first covers the broad subject of GIS and deals with applications of GIS to library and information science. Finally, the search pertains to identifying literature concerned directly or indirectly with the research topic.

After considering the impact of GIS on library and information field, the third chapter attempts to outline the various possible uses of GIS in the library environment. The following uses of GIS in the library are discussed in detail.

(i) Use of GIS in mapping of library collection.
(ii) Use of GIS in providing information services based on spatially referenced databases.
(iii) Use of GIS in mapping/networking of library resources.

The first use is, in fact, discussed in more detail in the fourth chapter and as such the third chapter concentrates on the second and third uses. This chapter has also touched upon all other uses of GIS in the library.

As stated above, the fourth chapter, concentrates on "Use of GIS in Mapping Collection". The discussion begins with issues like the emergence of OPAC and its developments in different generations. It further discusses Swanson's vision about the future of library catalogue. The study then discusses theoretically the concept of the mapping of library collection by using GIS. The discussion refers to various issues like the presentation of location details in OPAC, significance of location of library material, role of library classification in subject grouping as well as in retrieving the library material, need of location finding mechanism from the users' point of view, need of handling locations data separately and finally mapping the library collection using GIS with its prospects and advantages. The study has considered the situation of immobile library shelving as a basis behind the idea of mapping of collection. However, the static library environment or the store house situation has been used as the starting point as one of the developmental stages in presenting the further stage. While doing so, the study discusses, perhaps for the first time, modelling the shelving process of the library by introducing idea of "shelving model", which can monitor the shelving activity of library collection and give guidance to library staff. The chapter also discusses other related issues like virtual browsing and comparison of virtual library and simulated or GIS based library systems.

Due to non-availability of relevant literature on the topic and due to lack of systems based on the ideas discussed in the preceding chapter, it was felt necessary to demonstrate at least one of the important concepts viz. "Mapping of Library Collection Using GIS". The fifth chapter, therefore, discusses the development of the prototype in terms of its various aspects. They include : identifying a suitable library, preparing floor diagrams with necessary location features drawn in them and identifying suitable GIS packages. The chapter further deals with the actual steps of prototype development using GRAM package, such as digitizing the library floor maps, creating attribute data, establishing links between them and finally demonstrating the concept of "Mapping of Library Collection" and retrieving location of library material in a digitized map on computer screen. The results are provided (in Appendix II) in the form of photographs of computer screen outputs.

The sixth and final chapter presents the conclusion of the study, the research findings, recommendations and conclusions drawn at the end of the work. In addition to prototype results other appendices enclosed pertain to information about GIS technology and the exhaustive bibliography comprising of references directly used in the study.

The keywords which were identified during the course of the study are listed below :

Classification—Subject Grouping Tool
Classification—Location Indexing Tool
GIS Based Library Systems
Geographic Information Systems
Geographical Information Systems
Geographic Mapping Subject Contents
Library Inventory Management Tools
Library Networking
Library Shelving Model
Location Data
Mapping Library Collection
OPAC
Simulated Library System

Spatially Referenced Databases
Thematic Mapping
User Study
User Education
Virtual Browsing
Virtual Library
Virtual Library Network

ACKNOWLEDGEMENTS

I would like to thank my guide Prof. S.R. Ganpule for inspiring me to undertake the research, for keeping my morale and sustaining myself in the research process and for his role as a colleague and guide.

I would also like to thank authorities and staff in Centre of Studies in Resources Engineering (CSRE) of Indian Institute of Technology, Bombay for their kind and generous support in preparing the prototype. Special thanks are due to Mr. Somanchi Chandrasekharam.

I take this opportunity to thank Prof. M.R. Riswadkar for his valuable suggestions and interest shown in the work. I am extremely thankful to Dr. Nandita Bose and Prof. M.S. Malshe for their valuable suggestions while finalizing the manuscript.

Thanks are due to all those who have supported me in particular Mrs. Sonia Relan (DESIDOC, New Delhi), Mrs. Bhavana Lizara (Overseas Press, Delhi), and Mrs. Pratibha Dadarkar and all my colleagues, friends and staff of Central Library of IIT Bombay, and I am extremely thankful to the authorities of University of Pune for permitting to publish the research work appropriately.

Finally, I thank Mr. S.P. Sinha of Concept Publishing Co. for accepting difficult task of publishing the research work and Mr. Bindu Juwale of Universal Book Service, Pune for his help in this venture.

I cannot forget the patience and the sense of responsibility shown, in sacrificing their joyful time, by my family members Mrs. Kumud, Miss Deepali and Master Saurabh.

This work was possible mainly because of my father's faith in the intelligence of his ordinary son and due to blessings of my spiritual Gurus.

D.N. Phadke

CONTENTS

LIST OF FIGURES AND TABLES

Figures

Tables

LIST OF PLATES

1

Introduction

WHAT IS GIS?

Traditionally, geographic information has been represented in terms of maps, charts, atlases, earth globes etc. While looking at maps, our questions are answered by establishing geographic relationship with them. We plan our trips, tours, hikes and decide locations, places of visit on the basis of maps and charts. In the process our planet is perceived and understood by using maps. Causes of happenings like earthquakes and volcanoes are also understood through maps and the information incorporated in them.

Thus, Geographical Information Systems have been in use for a long time. What we were doing earlier, is being optimized/utilized more efficiently today due to technological developments, especially those in computer technology. As Cox describes[1] "We are all walking Geographical Information systems... In essence a GIS is old automated GIS is new".

Paper maps, traditionally consulted for geographic information, have several shortcomings as they are difficult to update, manipulate and combine with other data in order to facilitate analysis and interpretation. Computerization of maps has overcome all these shortcomings. It has greatly improved the usability of geographically referenced information resulting in the emergence of a new field called "Geographical Information Management (GIM). GIM has two components namely (a) the activities of users and policy or decision makers and (b) technology viz., Geographical Information Systems (GISs).[2] GIS takes care of computerized mapping and related geographic information processing. Like GIS there are other technological systems used for automated mapping viz., Computer Aided Mapping (CAM), Computer Aided Design (CAD), and Automated Mapping and Facilities Mapping (AM/FM).[3] At this juncture it is essential to elaborate on the genesis, nature and functions of GIS.

GIS : Conceptual Background

Definitions

A most comprehensive treatise, *Geographical Information Systems: Principles and Applications* ed. by Maguire, Goodchild and Rhind,[4] wherein authors give a dozen definitions of GIS, is useful for a basic understanding of the concept. The other principal source is the book, *Principles of Geographical Information Systems,* for *Land Resources Assessment* by Burrough[5]. Burrough describes the GIS as a system that attempts, "to develop a powerful set of tool for collecting, storing, retrieving at will, transforming and displaying spatial data from the real world for a particular set of purpose. This set of tools constitutes a Geographical Information System' (Sometimes referred to as a Geographical Information System). Geographical data describe objects from the real world in terms of (a) their position with respect to a known co-ordinate system, (b) their attributes that are unrelated to position (such as colour, cost, pH, incidence of disease etc. and (c) their spatial interrelations with each other (topological relations), which describe how they are linked together or how can travel between them".

Rowley[6] has formally defined GIS as "A system for the capturing, storing, checking, integrating, manipulating, analysing and display of data which are spatially referenced to the earth. This is normally considered to involve a spatially referenced computer database and appropriate application (GIS) software".

Smith and others[7] have defined GIS as "a database system in which most of the data are spatially indexed and upon which a set of procedures operate in order to answer queries about spatial entities in the database". In Congalton and Green[8] words, GIS is "a system for entering, storing, manipulation, analysing and displaying geographic or spatial data. This data are represented by points, lines and polygons along with their associated attributes". Franklin[9] concisely defined GIS as "a computer system designed to collect, store, retrieve, manipulate and display spatial data".

Geographical Information Systems thus is a special kind of information system which gives a combined effect of digital cartography, CAD and efficient data management systems. GIS handles the spatial data with reference to its location in space.[10] As stated by Franklin[11] "the GIS represents a new category of software and an excellent practical application of hypertext/hyper media. That is, most GIS software uses map regions and symbols to hide detailed information that can be revealed at the touch of a button".

Historical Background

Roger Tomlinson and his co-workers developed the Canada Geographical Information Systems (CGIS), as a part of government initiative to inventory natural resources in the southern of Canada in 1960s and this was the origin of today's GIS. Maps produced by CGIS were used to rehabilitate marginal farmland and manage timber production. At the same time Harvard University's Laboratory for Computer Graphics and Spatial Analysis devised a raster based computer mapping package called "SYMAP". In 1969 the Environmental Systems Research Institute (ESRI) was founded at Redlands, California. ESRI produced raster based package called "GRID". Hereafter the researchers at Harvard Lab and ESRI have progressively developed and streamlined techniques for newer and improved versions of GIS. In 1980s ESRI brought out the successor of GRID, namely ARC/INFO, primarily a vector based GIS system. Since then GIS software developing firms have gathered momentum and a large number and variety of GIS softwares are continuously brought out since then. They range from mainframe to desktop and from Macintosh to IBM compatible. There are three major vendors of GIS producers today and they are ESRI, Intergraph and ERDAS. ESRIs' ARC/view is one of the most famous PC based GIS software packages available today.[12]

How GIS Works?

Necessary Support of GIS Operations

Geographical Information systems have three necessary support components:

- Computer Hardware,
- Sets of application software modules, and
- Organisational component.

Computer hardware of GIS comprises of the Central Processing Unit (CPU) linked to the storage unit which provides space for data and programmes. *A digitizer* and other peripherals such as digitizing board and input pointer are used to convert data from map and documents into digitized form and to download the same to the computer. *Plotter* and other display devices are used to present the results of data processing. *Tape drive* is needed to store data and programmes, and to communicate with other systems. Inter-computer communication within the network is made possible via *modem*. The user controls the computers and peripherals via Visual Display Unit (VDU). The user terminal itself may be a micro-computer.

Software package for GIS comprises of five basic technical modules :

- Data input and verification.
- Data storage and database management.
- Data output and presentation.
- Data Transformation.
- Interaction with the users.

Data input covers converting data from existing maps, fields, observations, and sensors (from satellites or aerial photographs) into compatible digital form.

Data storage and database management deal with the way in which data about position, linkages (topology) and attributes of geographical elements (points, lines and areas) are structured and organised. Computer programmes that organise such data are known as Database Management Systems (DBMS).

Data output and presentation involves the mode in which results of analysis are displayed.

Data transformation focuses on the maintenance of data, making it up-to-date, matching it with other data sets and removing errors.

Interaction with users involves facilitating users to interact with systems directly in a simple way and to get the output in a manner can be easily interpreted.

The organisation aspect in the use of GIS relies on organisational support such as adequate hardware, competent GIS package to handle complexity of problems and to provide training for the system users. Only then is the investment worthwhile and fruitful, and the system efficient in meeting its purpose.[13]

Data Elements and Types of Data

GIS handles two kinds of data (i) geographic or coordinate data called *Spatial data* and (ii) *attribute data.* Spatial data represents features which have known locations on the earth. While working with maps, spatial data is recorded and classified according to three sets of coordinates viz., point (one dimension), line (one dimension) and area (two dimensions). The information relating to physical and spatial features referring to point, line or an area is stored on the basis of their respective coordinates in the computer. Attribute data are the information linked to the geographic features (spatial data) that describes the features. It is non-geographic information associated with point, line and area elements in GIS. It is often called '*Z*' value in GIS. Spatial data or location is described in terms of *x* and *y* coordinates and *z* value is attribute information. For example data relating to highway is stored in GIS, where spatial data contain a series of latitudes and longitudes of highways in *x* and *y* coordinates that represents a highway on the two dimensional map, whereas corresponding attribute data about the highway will have information such as the type of pavement, route, road name, the number of lanes it comes in contact with and the historic accident information.

In addition to these data requirements viz., spatial and attribute data, the GIS must have databases in data layers i.e., the data should have topology and the system should be capable of overlaying the data sets with a view to manipulate spatial data in order to derive new data or relationships.

Data layers : Here GIS is capable of creating data layers of digital spatial data. A data layer consists of a set of logically related geographic features and their attributes. A large scale topographic map usually contains vast amounts of information about geographic features and their relationships. In order to maintain the accuracy of geographical features, it is necessary to have structural models of the entire map so that it is possible to split the information into different structures or theme based maps/layers as and when required. Simultaneously, one can have a full-fledged map. GIS makes this possible. This structuring or layering facility is used to provide a variety of combination or overlaying layers to create new maps or structures/layers.

Topology : Topology is the "way in which geographical elements are linked together".[14] Topology describes the links and relations among geographical features. It is built on two concepts viz., absolute location and relative location. The absolute location feature of an entity is represented in terms of co-ordinates (x and y) whereas relative location is described as the location of that entity in real life. For example the location of University of Pune in terms of absolute location will be A^o East longitude and B^o North latitude. However, its relative location will be that it is within the state of Maharashtra, '*n*' kilometres from Mumbai, in Pune district, north-east of Pune at Ganesh Khind. As described by the Cox[15] "Topology is the critical element that distinguishes a GIS from graphics or automated cartography systems".

Overlaying Geographical Data : In overlaying function, various data sets are stacked one above the other and viewed as if on a light table having some common projection or same coordinate system. This is done for various kinds of analyses and experiments to create a new map as per need. The ability to overlay spatial data sets is one of the capabilities that GIS provides to the users with its analytical power viz., integration of different kinds of data.

Vector and Raster Data Formats : In GIS data is represented in two data formats called raster or grid structure and vector data formats. Raster format is the older form and it is stored in a grid or pixel format which refers to some coordinate system (latitude and longitude).

The size of the grid varies, and determines spatial resolution. Raster data is easily manipulated by the computer but requires more storage space. Digital remotely sensed data or satellite imagery data is an example of raster of grid data. In Vector-based data format, geographic features are represented in a series of coordinates. Vector type data requires comparatively less space in the computer and it is preferred for displaying a true shape of the object which is not possible in raster or grid type of data format.[16]

Applications of GIS

As GIS has special characteristics, which differentiate it from other information systems of handling spatial information i.e. information referenced by its location in space and because geographical location is a central consideration in the planning process of many activities, GIS finds its applications primarily as the decision support tool useful where spatially referenced data is considered in the decision making process. Traditionally the areas of application of GIS have been :

- Resource Management

- Land Use Planning
- Defence
- Utilities and Environmental Monitoring
- Demographic Analysis

and have widened unlimitedly in the recent years to numerous non-traditional areas too.[17] GIS technology has provided large scale applications in the traditional areas and is now available on smaller hardware platforms. This adaptation of GIS technology to a variety of demands has facilitated its use in wide ranging forms depending on the complexity of the given problem. The application of GIS in the above areas are well established. The last two or three years have witnessed increasing applications of GIS in business planning and even marketing. As indicated by Franklin[18] as much as 80 per cent of all information stored by business and government may be geographically referenced. Examples of such information are: Census data, mailing addresses data, facility layouts, telephone numbers data, land-use information and details about area networks such as water, electric, gas, sewerage, cable, transportation and telephone systems. Other business applications would include marketing, site evaluation, sales and competition impacts, forecasting, merchandising, media and recruitment planning, customer satisfaction,[19] and also warehouse monitoring and inventory management. Applications of GIS are growing in innumerable directions and it is difficult to make an exhaustive account of it.

LOCATION CONCEPT IN LIBRARY

Traditionally, libraries have been using classification schemes for classifying library material in order to have some definite order, called "classified order" in arranging reading material on the library shelves. The library system provides each book a "call number" comprising of classification number, accession number and author mark. The book, indicating the call number on its spine is kept on the shelf at an appropriate place in the classified sequence. The corresponding catalogue data is generated manually or in computerised form and the same call number is also incorporated in the catalogue data. It is expected that the user, after consulting catalogue or OPAC, would retrieve the call number indicated and he/she would use the same in retrieving the book from the shelf.

Classified arrangement also facilitates browsing of books on allied topics on library shelves and such browsing has many advantages in getting the reading material in the library by the readers.

Any classification scheme is expected to perform the following functions in the library as described by Line[20] :

1. "To arrange books in such a way that some kind of direct subject searching and browsing is possible.
2. To be able to retrieve material on a subject with reasonable precision.
3. If the library has catalogue, it usually utilizes class numbers to guide the user to the right place on the shelf".

Thus classification system is also supposed to guide the users in finding the place of the reading material on the shelf, whereas a catalogue performs the role of linking catalogue data to classified arrangement of books, through the call number. The reader uses the call number retrieved after consulting the catalogue to locate the book in the library. However, the structure of the call number in no way provides any indication of the actual location features of the material in the library. There is no other device used in the library to interpret the call number in terms of location of that call number in the library premises. Neither the computerised catalogue nor the library staff constantly provide assistance to the users in this regard. Thus the location concept is handled very roughly or arbitrarily by library systems.

Types of Location Data in the Library

There are various types of location data associated with a library and the collection of reading material in it. The library itself has a geographical location within the institution, which may be on a campus. If there are different subject or departmental libraries, each of them will have particular geographical location details. If a library is a multistoreyed building, each floor number becomes an additional location feature. Due to form type and other status details, library materials are kept at different locations in the library such as:

- General Collection
- Reference Collection

- Special Collection (such as MSS or personal donations)
- Classified Collection
- Rare Collection
- Non Book Materials
- Microforms/microfilms Collections
- CD Collection
- Compact Storage Collection
- Current Journals
- Back Volumes
- Text Book Collection
- Book Bank Collection
- Video Collections
- Thesis Collection
- Standards Collection
- Patents Collection
- Reports Collection
- Abstracting or Indexing Source Collection
- Newspaper Collection
- Map or Atlas Collection
- Magnetic Media Collection, etc.

Each item added to the library collection, because of its form, type or convention, goes to one of the above collections, which are kept separately. Conversely, all materials on one subject also goes to different places as there is no other system in the library to keep the materials on one subject at one place irrespective of their forms. Over and above, each item or book has a unique location in terms of its place in a stack room, in a row, in a rack or on a partial shelf in the library premises. As such the library has a large locational data.

Current Location Data Handling Systems

It is seen that classification schemes help in arranging books on shelves in such a way that the arrangement provides some kind of direct subject access and facilitates browsing. At the same time it tries to provide guidance to the readers through the library catalogue to find the place of reading material in the library. However, in reality classification is primarily used for arranging books on the shelves providing some kind of subject access and facilitating browsing. It is further emphasised that classification scheme does its best in providing guidance to the users to understand the whereabouts of the library material but it is essentially not the primary function but only an added advantage. Classification scheme can become a true subject access tool but it cannot also become a location index. This is because, providing location and also providing classification do not go together. There are several practical difficulties in using class number as call number or location number. The view of using class number as location number is predominant because in the library too many restrictions have been put on classification work. Classification schemes are continuously developing and thereby they are kept up-to-date. In practical classification, however, the library uses only an abstract form of it. More often than not an age old edition of the classification scheme is used for classification. This is necessary in the given situation as it is impossible to maintain the classification scheme up-to-date in practice, but it would create a huge reclassification and recataloguing work. The Library, therefore, is not in a position to adopt each and every revision of the scheme. And it is a fact that if classification scheme is practiced in its true form, class number cannot become the location number.

Now with the availability of computers, the situation has changed. Library can maintain the precise and up-to-date classification scheme without much efforts. Precise and updated classification would definitely enhance subject access and make browsing more meaningful.

Problems

The opinion that the library should use classification scheme to classify its materials is not being challenged. However, in the process of organising the collection in the library, material on the same subject gets scattered because they represent different forms and types. From the library management point of view it is convenient to keep the like material together.

This is particularly true in the case of secondary and tertiary type of material. In the case of the primary collection of the library, users do not get the exact location of the books from the call number. No other system exists in the library which provides continuous guidance to the users in this regard. Library staff is also inadequate in number to help the readers.

These complications are created by the library system which is a necessary evil. It is the fact that as a result of organisation of material in the library it gets scattered and several difficulties are faced by the readers in understanding what is stacked where. Every time, the user has to interpret the special codes that invariably appear as part of call number in order to find the actual location within the library. As stated by Stephens[21] "Library OPAC displays have not done much to develop location (handling system)...a truly "Friendly" system should be expected to spell out location plainly. In fact because library shelving is quite immobile, it is not unreasonable to expect OPAC displays to show the precise location of each item in the collection, down to the specific range, stack, and the shelf number". The situation further becomes worse if the library uses a classification scheme which makes use of various symbols in number building. To know context, meaning and sequence of various symbols while interpreting class numbers is really bothersome for the users. The users need to get the book physically in a short time without going through complicated procedures.

The Reference Librarian of Wayne State University's Purdue Library Mr. Kane,[22] who invented Hypercard call number directory, also has expressed his concern about the problem faced by the library users in finding reading material. He states that "readers face with the general question of How do I find the books by call number in this library". The response which he experienced while implementing Hyper-card call number directory systems in the library confirms that readers do require a short cut method in finding locations of the library material. Prof Grimshaw[23] has described the problem in his words as "Existing library information systems are good at answering questions 'What', yet neglect the important question of 'Where' In some sense the where question is answered by providing some general locational reference such as 'main campus library—3rd floor'. Mapping of such data, perhaps to a more specific location, is possible using GIS and could provide the information scientist with a useful inventory and facilities management tool".

RELEVANCE OF GIS IN THE LIBRARY

GIS is a special kind of information system which efficiently manages databases and handles data including the locational or geographic dimension. The application area of Geographical Information System is very wide. It has emerged from the domains of Land Use,

Demographic and Environmental applications. GIS technology has proved its large scale applications and is now available on smaller hardware platforms with its capacity to map data on a desktop system. Thus, GIS has opened the gateway to a variety of applications in various fields. GIS has already influenced and made impact on the field of library and information science as well, in many ways. Firstly, impact is seen in the availability of voluminous information on GIS, the task before the librarians now is to handle this vast amount of information. Secondly, due to the variety of applications of GIS in different fields, users of geographic information are putting pressure on the library as the central place for information on GIS, information services based on spatially referenced data and knowing the use of GIS technology. Thirdly, GIS is impacting on librarianship, not only participating library as intermediator in providing such services, but also in using GIS as one of the tools in library management.

Applications of GIS in Library

Geographical Information System is being used as an inventory management tool in various fields, such as mailing addresses or telephone numbers. It is also used in networking areas such as water, electric, gas, sewer, cable, transportation and telephone networks, and also for navigation. It is possible to draw a map showing the location of a road. It is also possible to store details relating to each road such as its name, its width, house numbers and details of persons staying in those houses. Systems like this can be developed for a township and one can find the location of person staying on a particular road, in a particular house, or on a particular floor of the house. The data and the system can be used to perform a variety of analyses. After seeing these types of GIS applications this study was undertaken to explore the use of GIS in the library. Handling of locational data in the library has become necessary because library management cannot depend on the locational data handling capacity of classification system which is not the primary function of the scheme. Presently such data is handled very arbitrarily which is not conducive to the satisfaction of the users.

It is seen in the earlier discussion that library has a good amount of location data. The question is whether GIS could be of any use in handling the locational data of library materials. The present topic of discussion viz., the virtual reality library systems which appear quite frequently in library literature is really a simulated library system. In the course of this project, the researcher had the opportunity to attend a short course on thematic mapping using GIS. This was utilised to try to digitize a small diagram of library stacks to answer a few location related questions. This small experiment led to the conviction that it would be quite useful to utilize GIS in the library. This idea was further discussed with some experts in the field of GIS technology who gave positive feedback. All these efforts revealed that GIS could be used in the library, especially in handling locational data of the library material. The initial step was to study Use of GIS in mapping library collection.

Computers are already used in the library for various housekeeping functions. The age old card catalogue has been converted into an electronic form called On-line Public Access Catalogue (OPAC). OPAC is under continuous development and newer and newer versions are made available for the benefit of the users. It was possible to provide circulation status of a book in OPAC while referring to the bibliographical details of the same. Some people have tried to provide broad location statements in the OPAC display.[24] However, it was yet to incorporate the exact location of the library material in the library through OPAC though such information is necessary.

Mapping of library material using GIS was altogether a new idea. GIS makes it possible to draw floor diagrams of various floors of the library building, along with furniture and fixture items kept there and to convert the map into digitized form. Each book in the library can be given location details and then the entire bibliographical data can be linked to the floor maps through location codes. The system would then be able to show the simulated view of the library floors on the computer screen.

The system could be used to answer many queries relating to finding the location of specific library material. The system thus could provide an inventory management tool and could enhance the capability of OPAC.

As GIS is being used for the networking of various services, similarly it can be used in planning and establishing library networks. Libraries could be mapped using a suitable map, location and other details. It could be used in preparing models of library network using above data and in providing feasibility results which could be used in actual planning of library networks.

It is observed that libraries are trying to use GIS in providing library services using spatially referenced databases especially in using census data. The idea could be further explored in using other useful databases and providing library services to the users.

If the idea of "Mapping of Library Collection" is found to be successful it can also be used to provide library instruction to the users and will also help in incorporating users education in classroom teaching. Use of the simulated view of the library premises, its functional locations and details of services could be successfully done through such programmes. In addition to the above, GIS might find its applications such as (i) "Use of GIS in mapping subject contents against geographical treatment given by classification schemes", (ii) Use of GIS in user study, and (iii) Use of GIS in mapping and networking of book suppliers and distributors also.

Benefits

The use of GIS in the library as proposed in the present research work is likely to offer several benefits to the library management and to the users, such as :

1. It can help users in finding exact location of the library material thereby saving their time.
2. Users can visualize the availability of material in various forms and kept at different places in the library just by sitting across the computer terminal
3. Conducting survey for proposal of library network by mapping library locations would be easier for agencies like Library Associations; who are primarily concerned with such projects.
4. Library can be looked upon as source for spatially reference information.
5. Users can get knowledge of the use of GIS technology in the library.
6. Different kinds of subject mapping could be done by exploiting subject treatment given in the classifications which is normally not possible in classified catalogue.
7. User surveys could be conducted with a view to exploit the potential users which are geographically within the reach of the library. The surveys can be used to analyse the users data in various context and the results could be used in decision making.

8. Similar to library networking, the book selling and supplying activity can also be coordinated by using networking and mapping techniques. The result could be used in preparing models for book purchasing activity of a library or for group of libraries.
9. Browsing books in terms of bibliographic and other data especially in the absence of books on shelves.
10. It would be possible to see the enhanced details of books especially non-conventional details like content page index, and images of various parts of books and to know details of interdisciplinary subjects.
11. Having comprehensive collection on the library database, it would be possible to see literature on the subjects available in various forms and kept at different locations in the library through a simulated view of the library collection map.

AIMS AND OBJECTIVES OF THE RESEARCH WORK

The aim of the present research work is to study and try to explore the possibilities of use of GIS in library in different ways in general, and to concentrate on following three applications of GIS in the library, in particular :

(a) Use of GIS in Mapping the Library Collections..
(b) Use of GIS in providing information services based on use of GIS and spatially related databases.
(c) Use of GIS in Mapping/Networking Library Resources.

Plan of the Research Work

At the outset it is proposed to go into extensive literature survey, on the topic by referring primary, secondary and tertiary sources of the field of library and information science. The topic being of interdisciplinary nature it is necessary to seek references from other fields like Geographical Information Systems and Computers also. It is also necessary to have a clear understanding of the basic details of GIS, because it is a relatively new and unconventional field of knowledge. The study noted down the various facets of the topic and accordingly, the organizational plan of the thesis has been prepared. The organisations and experts in the field were contacted in order to discuss the topic with them and have their opinions on the research work. Accordingly a seminar was presented on the topic in the presence of group of persons including the guide, senior professors in the field of library science and the experts in the field of Geographical Information Systems and Library Automation with a view to assess and confirm the research proposal. Researcher attended a short course related to the applications of GIS. Relevant literature on research technique was studied and the research methodology to be used, was finalised.

The utilization of GIS in the library is an innovation with no precedents, and thus it was felt that due to non-availability of literature in support of ideas presented in the theoretical discussion, it would be necessary to conduct an experiment by preparing a prototype, and as such, a prototype has been designed. The newly discussed concept was demonstrated through the results of the prototype.

RESEARCH METHODOLOGY

Type of Research

The present study would be an exploratory type of research. It is a preliminary investigation into various uses of GIS in library. It attempts to find out whether the use of GIS has any advantages over the present systems. Having done this and getting a positive response, further study may be planned to develop the system, and to assess its performance.

Assumptions and Requirements

The following assumptions and requirements were considered :

(i) Library shelving is quite immobile from practical point of view.

(ii) Libraries in future will have GIS technology in the form of suitable GIS software or have access to the GIS systems.

(iii) Library will receive in its collection spatially referenced databases.

(iv) Libraries agree to the idea of resources sharing and putting their catalogue and bibliographic data on the internet/www.

(v) Library database will have comprehensive collection with non-conventional details, including locational data.

(vi) Library would prepare a systematic shelving model and use it in monitoring shelving activity.

Developing a Prototype

The initial literature survey revealed that one of the facets of the topic viz., Mapping of Library Collection was altogether a very new subject and it was unlikely that the researcher would get relevant literature or the information of availability of similar systems existing elsewhere. It was, therefore, decided to have combination of theoretical discussion as well as experimental backup with a view to provide support to the introduction of this new idea.

The literature survey is presented in the next chapter.

NOTES*

1. Allan B. Cox, "An Overview of Geographical Information Systems" *Journal of Academic Librarianship,* 21, No. 4 (July 1995) : 237.
2. Carl Franklin, "An Introduction to Geographic Information Systems: Linking Maps to Databases" *Database* 15, No. 2 (April 1992) : 12.
3. *Ibid.*, : 13.
4. David, J. Maguire, Michael F. Goodchild and David, W. Rhind, *Geographic Information Systems : Principles and Applications*, 2 Vols. (Marlow Longman Scientific & Technical; 1991).
5. P.A. Burrough, *Principles of Geographic Information Systems for Land Resources Assessment* (Oxford : Clarendon Press; 1990), 6.
6. Jennifer Rowley, "The Nature and Applications of Geographical Information Systems", *ASLIB Managing Information* 14 (April 1994) : 21.
7. Maguire, Goodchild and Rhind, *op. cit.*
8. Russell G. Congalton and Kass Green, "The ABCs of GIS : An Introduction to Geographical Information Systems", in *Wetland and Environmental Applications of GIS,* ed. by John B. Lyon and McCarthy (New York : CRC Press Inc., 1995), 9.
9. Franklin, *op. cit.*, p. 13.
10. Rowley, *op. cit.*, p. 20.
11. Franklin, *op. cit.*, p. 13.
12. *Ibid.*
13. Borrough, *op. cit.*, p. 6-10.
14. *Ibid.*, p. 183.
15. Cox, *op. cit.*, p. 240.
16. *Ibid.*, pp. 238-239.

* Reference are cited as per *The Chicago Manual of Style* 4th edn. (Chicago University of Chicago; 1993).

17. Rowley, *op. cit.*, p. 21.
18. Franklin, *op. cit.*, p. 12
19. Rowley, *op. cit.*, p. 22.
20. Maurice B. Line *et al.*, "How Golden is your retriever? . Thoughts on Library Classification" in *Lines of Thought : Selected Papers of Maurice B. Line*, Ed. by L.J. Anthony (London : Clive Bingley, 1988), 13.
21. Irving E. Stephens, "Getting more out of call numbers: Displaying holdings, locations and circulation status", *Cataloguing and Classification Quarterly,* 13, No. 3/4 (1991), 99.
22. William P. Kane, "A Hyper-card call number Directory. Using stacks to finds stacks", *College and Research Library News,* 50, No. 7 (July-August 1989): 577.
23. David J. Grimshaw, "Geographic Information Systems in Business" In *Encyclopedia of Library and Information Science,* Vol. 54, Ed. by L.N. Kent (New York : Marcel Dekker, 1994), 198.
24. Stephens, *op. cit.*, p. 99.

2

Review of Literature

INFORMATION SOURCES USED

Although the concept of using GIS in library is a new phenomenon, it was found necessary to check whether there would be some literature available pertaining to the topic. With this expectation in mind the following sources were tapped :

- Library and Information Science Abstract (LISA).
- Information Science Abstract.
- Advances in Librarianship (Academic Press).
- Annual Review of Information Science and Technology (American Society for Information Science and Learned Information INC).
- Current Awareness Abstracts (ASLIB).
- Encyclopaedia of Library and Information Science (Marcel Dekker).
- Geographical Information Systems: Principles and Application ed. by Maguire, Goodchild and Rhind.[1]
- Classification of Geographical Information Systems Literature and Application.[2]
- International Journal of Geographical Information System (Pegamon Press).[3]
- Bookscope.[4]
- Several other library and information science journals.

Several difficulties were faced in having access to the important abstracting and indexing journal LISA as holdings of its back volumes were incomplete in University of Mumbai Library on account of its discontinuation due to shortage of funds. There was no source available in Mumbai to access LISA on CD-ROM. It was, therefore, decided to go out of Mumbai to complete the task of literature search. Two visits were made to Delhi in this connection to get the basic source of information.

Organisations in Delhi like Indian National Science Documentation Centre (INSDOC), Defence Science Information and Documentation Centre (DESIDOC) and India International Centre (being Headquarters of Delhi Library Network) were personally visited. Assistance was also sought from Indian Institute of Science, Bangalore, Informatics, Bangalore and Systems Research Institute, Pune to get relevant reference.

ORGANISATIONS CONTACTED

As a part of literature search correspondence was made with following organisations and institutions situated in the country and abroad with a view to pursue the above objectives :

- Environmental Systems Research Institute Inc (ESRI) California, (Mr. Dave Buyers)
- Association of Research Libraries, Washington, (Mr. P.S. Adlers)
- National Centre for Geographical Information and Analysis (NCGIA), University of California, Santa Barbara, (Mr. Michael F. Goodchild)
- Warwick Business School, University of Warwick, Coventry (UK) (Prof. Grimshaw)
- Purdue Library, Wayne State University, Detroit, (Mr. William P. Kane)

Attempt was also made to flash the news about the on going research work in the Bulletin Board of GIS user group through Ernet/ Internet facility.

OBJECTIVES OF LITERATURE SEARCH

The literature search was conducted with the following objectives in mind :

1. To study the basics relating to Geographical Information Systems (GIS) and its applications in various fields.

2. To find out the references published, in a broader perspective, which are indirectly related to the topic.
3. To find out the evidences, if any, about existence of GIS based library system at elementary, experimental or advanced stages.
4. To trace the literature in various sources which is directly relevant to the topic of the research.

REVIEW OF LITERATURE

Basic Literature on GIS

GIS Technology has emerged only in recent years but since then it has grown speedily. It is seen that many countries have been using GIS and have contributed to the research relating to GIS while organising information about their geographical sources. Among those, countries in the British Commonwealth and in the European continent, seem to have contributed more share to world class GIS research due to their long tradition of geographic study.

Cartographic societies from various countries have published articles on GIS in their journals. They include Canadian Cartographic Society, British Cartographic Society, Australian Institute of Cartographers and American Congress on Surveying and Mapping.[5]

Currently there are good number of journals that publish literature on GIS. They include :

- Cartography and Geographical Information Systems (Earlier named American Cartographer).
- International Journal of Geographical Information Systems
- Journal of Photogrametric Engineering and Remote Sensing.
- Geo Info Systems.
- GIS World (monthly) etc.

In addition to these there are some more journals that publish articles on GIS. All of them are not scholarly journals. *International Journal of Geographical Information Systems* (which began in 1987 as a quarterly, now is bimonthly) is one of the principal scholarly journals. *The Annual Geographical Information : The Year Book of Association of Geographical Information,* contains many scholarly and descriptive articles on several GIS topics plus the trade directory (For British Market). GIS World is the principal trade journal. There are many newsletters published by professional associations that compile and publish details of the organisational activities on the subject, for example *URISA News Letter.*[6]

Several monographs devoted to GIS have been published. Many publishers have concentrated on GIS literature by bringing out monograph series. Academic Press has published a monograph viz., *Fundamentals of Spatial Information Systems* as a part of its APIC series. Oxford University Press has a series called *Spatial Information Systems*. The recent title published under this series is *Mapping Spatial Information Systems.* Pergamon Press publishes a series under the title *Modern Cartography.* It has brought out two titles under this series viz., *Geographical Information System : The Micro Computer and Modern Cartography and Visualization in Modern Cartography.*[7]

There is only one comprehensive two volume monograph on GIS to date, viz., *Geographical Information Systems: Principles and Applications* ed. by Maguire, Goodchild and Rhind.[8] The National Centre for Geographical Information and Analysis (NCGIA) publishes a very useful technical paper series which concentrates on current research and publications on GIS.

GIS World Source Book (Annual) is the comprehensive source of information which covers virtually all aspects of GIS.[9] As far as GIS terminology is concerned a very good source called *Practical Hand Book of Digital Mapping : Terms and Concepts* is available.[10]

A book by P.A. Burrough[11] entitled, *Principles of Geographical Information Systems for Land Resources Assessment*, published by Oxford University Press is the fundamental textbook which cannot be missed by researchers.

Because of the interdisciplinary nature of GIS and its varied applications, many articles are being published in several non-traditional areas also, making it difficult to have bibliographic control. This is because of concurrent applications of GIS in different areas. Most of the literature is available in the conferences and seminar proceedings where the applications of technology are reported first. The earliest among these was the *Auto Carto Annual Conferences* in 1973.[12] The Auto Carto has now been merged into GIS/LIS Proceedings and URISA (URISA Proceedings : Papers from the annual conference on Urban and Regional Information Systems Association). The other principal conference is AM/FM Conference. The 32nd Annual Clinic on Library Applications on Data Processing has been published under title *GIS and Libraries, Patrons, Maps and Spatial*

Information.[13] In India *XVII INCA Intl. Congress on Cartography and Administration* was held at Chandigarh in November 1997. ESRI/ ERDAS User Conference was held at New Delhi in December 1997.

There are several organisations and associations especially in the United States that have made efforts in generating literature, training users and promoting research in GIS. The roles played by Environmental Systems Research Institute Inc (ESRI), National Centre for Geographical Information and Analysis (NCGIA), Association of Research Library (ARL) and the Data User Services Division, (U.S.), Bureau of the Census are very significant.

Information on GIS is also available in CD-ROM form. A vast amount of information is available online on internet/www. There are three groups which discuss different aspects of GIS on internet. They are Geoweb, GIS-L and Maps-L.[14]

Thus information on GIS is developing with tremendous speed resulting in an ever-growing ocean of GIS information. This poses a problem for libraries who cater to the needs of GIS users, as processing and disseminating such a vast amount of information is extremely difficult. This review focusses on only that relevant part of literature pertaining to the topic. Our project deals with the application of GIS, so the methodology adopted is to select literature relating to basic ideas and concepts about working of GIS. This is found useful in understanding GIS application to library systems and in building future experiments on the GIS based library system.

Literature on GIS in Library and Information Science Literature

The fundamental and broad sources of information on GIS in various forms, give the basic idea and outline about the same. The next goal is to look for literature which is relevant and related to the specific topic of research i.e., the application of GIS in library and information science. Our (preliminary) survey proved that GIS has made a tremendous impact on the disciplines of library and information science.

Although the world of GIS is, at present, new to librarians as far as its practical applications in day-to-day library functions are concerned, it has tremendous potential to influence librarianship. A good number of articles on GIS are continuously being published in the field of library and information, and the trend is increasing. The following articles proved to be landmarks in the use of GIS in library and information management :

1. Article entitled "Geographical Information Systems in Business" by David J. Grimshaw[15] contributed in the *Encyclopedia of Library and Information Science,* Vol. 54 (1994).
2. The article published in *Annual Review of Information Science and Technology,* Vol. 27 (1992) by A. Shaffer[16] entitled "Data Representing for Geographical Information Systems".
3. Special issue on GIS brought out by the *Journal Information Technology and Libraries,* 14, No. 2 (June 1995), entitled "Making GIS a Part of Library Services".[17]
4. In the same year another special issue on GIS was brought out by *Journal of Academic Librarianship,* 21, No. 4, (July 1995) entitled, "Special Issue of Geographical Information Systems (GIS) and Academic Libraries".[18]

Literature on Application of GIS in Library and other areas

For convenience sake all the available literature was grouped as follows :

(a) Use of GIS in handling spatially referenced data in the library and in information services.
(b) GIS in Mapping of Library Collection based on location data.
(c) GIS in Mapping/Networking Library Resources.
(d) Use of GIS in other areas.
 (i) GIS in Business.
 (ii) GIS in Government Information in India.

There are several articles published pertaining to group referred as "A" which could be further divided into following two sub-groups :

1. Use of GIS in Map Librarianship and in handling traditional cartographic or map data.
2. Use of GIS in handling all other kinds of spatially referenced data.

GIS in handling Spatially Referenced Data

Spatial data could be defined as any item of information or activity corroborated with geographic features. Spatial data comprises of point, lines or area features. Such data may describe geology, soil, water distributions, weather, forests-cover, land-use, pollution, demographic and statistical features. Spatial data or spatially referenced data comes from maps, photographs, geo-coded data bases or from remotely sensed sources. There is a new trend that such data is generated in computer-readable form and compatible to GIS. In the light of these developments and on the background of growing demands from the readers for spatially referenced information, a new area of library services is being developed. A large group of articles have emphasised this particular emerging area viz., Library services based on spatially referenced data bases and GIS. In particular McGlamary and Lamont[19] have discussed the issue in addition to the use of Tiger Files; recommending upgraded information services such as virtual map library and the handling of spatially referenced data in general. It has also discussed a very prestigious project which librarians may find quite relevant, namely the ARL/GIS Literacy Project undertaken with the support of ESRI to increase awareness about GIS and the use of GIS technology among the librarian community.

Another interestingly titled article by North[20] viz., "Will your Library be the spatially data information centres of the future?", has visualized in a nutshell the future scenario about the library's role in giving information services using spatially referenced data, whereas the article entitled "Mapping 1990 Census Data" by Yott,[21] has discussed a case experiment where University of New Hampshire Library is actually engaged in giving such information to the public as per the requirements. The author has exhorted the librarian community to prepare themselves for future challenges in this regard.

Subsequently several articles have appeared which have discussed the preparation planning and implementation of spatially referenced information services in the library. An article by Franklin,[22] 'An Introduction to Geographical Information Systems : Linking maps to databases", has discussed in detail issues like hardware and software preparations required to initiate the services.

The special issue on geographic information systems, (GISs) and academic libraries published by *Journal of Academic Librarianship* edited by Peter Hernon has discussed the use of GIS in giving information services using spatially referenced data in detail right from an introduction to GIS to using Internet/www for this purpose. The fundamental features of GIS are discussed in the introduction by

Adler,[23] in this special issue. Cox,[24] has presented overview of GIS, while Abbott and Argentati,[25] have discussed GIS as a new component of public services. Hunt and Joselyn[26] have discussed ways to maximise accessibility to spatially referenced digital data. Longstreth,[27] has concentrated on GIS collection development, staffing and training. Cobb,[28] has discussed how conducive environment could be created to establish GIS based services. Stasser,[29] has discussed the use of DTP based GIS technology and its cost involvement. Allison,[30] has provided an example of GIS in use by describing how Automated Cartographic Information Centre (ACIC) of University of Minnesota's John R. Borchert Library made use of the Internet for giving map information using Electronic Atlas. Boisse and Larsgaard[31] have discussed the managerial perspective of the topic and lastly Cheverie,[32] has described the take off state of such services.

GIS and Map Librarianship

Computers have made it possible to handle cartographic data in a variety of ways, which otherwise was almost impossible manually. Application of GIS in this area has further given a boost to map librarianship. There are quite a good number of articles pertaining to this theme. Lang[33] in his article "Mapping the future of Map Librarianship" has discussed the use of electronic map and GIS in giving geographic information. Here map data source also include geographic data on CD-ROM. In USA nearly all depository libraries receive census data TIGER Files from US Bureau of Census on CD-ROMS. While using these sources the users are not only provided with extensive information along with maps but it has also facilitated users in creating their own maps as per their requirements.[34] The article by Allison[35] entitled, "University of Minnesota : Remote Access and the Internet" has discussed how libraries participating in ARL/GIS Literacy project jointly initiated information services on ARL Electronic atlas of USA through the Internet.

Another article by Lai and Gillies[36] viz., "Impact of Geographical Information Systems on the role of spatial data libraries", has also discussed this aspect in a wider perspective. The authors have drawn the attention of librarians towards the growing demands from the readers of information pertaining to the map data. Several articles which have discussed this aspect have been listed in the bibliography.

Use of GIS in Mapping Library Collection

A short write-up entitled "A Hyper Card Call Number Directory : Using stacks to find stacks" by Kane[37] was located where attempts have been made to devise a Hyper card system to handle location data of library material. The concern expressed for the acute need of location data in libraries was sufficiently close to our project and this article on the Hyper card system provided a view of an alternative system addressing this need. Another article entitled "Getting More out of Call Numbers : Displaying holdings, locations and circulation" status by Stephens[38] has discussed various means of handling locational aspects of library material broadly and the circulation status through OPAC.

Use of GIS in Mapping/Networking of Library Resources

Library literature does not accurately reflect the growing impact of GIS in mapping/networking of library resources. A stray article such as the one by Murr *et al.*,[39] studies the flow of information among the participating organisations in one of the information delivery networks.

There are a number of papers and articles referred to, some of them with titles resembling the topic of this research work which were found quite off the mark. They have mainly referred to the use of spatially referenced data and not the use of location data of the library material of the libraries.

Use of GIS in Other Areas

In the field of information, GIS is playing a very successful role especially in environmental management, land-use, urban planning, water resources management and remote sensing. Moreover, GIS has become interdisciplinary and no longer applied only to the above traditional areas. GIS is extensively used in many diverse areas also. An exhaustive report on the applications of GIS has been consolidated in Vol. 2 of the *Geographical Information Systems : Principles of Applications,* ed. by Maguire *et al.* There are number of articles published that describe varied applications of GIS. Diverse areas in which GIS is applied are listed below :

GIS in Business

The article, contributed by Grimshaw[40] in *Encyclopedia of Library and Information Science,* Vol. 54 (1994) viz., "Geographical Information Systems in Business", has discussed many possible applications of GIS in the business world. One such possibility is that one can link existing database to digital maps in order to explore the emerging relations which otherwise were not found previously. An article written by Rowley,[41] on "the nature and applications of Geographical Information Systems" has, in a general way, attempted to touch upon the wide range of different business application of GIS. Such as marketing, sales, competition impact, forecasting, merchandising, media and recruitment planning.

Select Application of GIS in India

Like in western countries, the influence of GIS is also seen in developing countries like India. There are several government and private organisations that are engaged in studying and using GIS technology for planning and decision making. They are also keen in creating awareness among scientists and engineers working in the concerned departments. In this case pioneering work is being done by NIC, DST, FSI, SOI, ISRO, SRI and CSRE IIT, Mumbai, and MMRDA, Mumbai. As special publication has recently been brought out by ISRO, Bangalore,[42] entitled "Organising Information Systems Around a GIS core [Database Organisation and Integrated Analysis using GIS (December 1994)]. A half yearly journal called *GIS India* is being brought under the sponsorship of Geo Map Society India, since 1992. Centre for Spatial Database Management and Solutions (CSDMS) is publishing a bimonthly journal named *GIS@Development.*

Other Peripheral Literature

Poulter has contributed two papers in succession viz., "Virtual Reality Library",[43] and "Building a Browsable Virtual Reality Library",[44] on a very new concept of "Virtual Reality Librar". A Virtual Reality Library depends on the use of OPAC by the reader wherein he/she is free to build up a conceptual model based on idea in his/her mind in such a way that imaginary compartments are created to browse through

OPAC catalogue on the same lines as books are browsed through on the shelves of the library. Though the system is interesting it gives just a feel of the real situation through a special kind of OPAC and does not discuss anything regarding presenting the real picture of the location of reading material in the library. The concept of "Virtual Reality Library" can itself become a full-fledged topic of research but it is in no way concerns the current topic of research.

CONCLUSION

The review of literature available on GIS and related areas, thus reveals that there is sufficient awareness of basic GIS technology. The literature provides clarification on the basic steps one should take before any application aspect is thought over. It is further useful in designing the plan for developing the prototype. The review of literature thus has helped the researcher in more than one way.

It was also possible to have sufficient information on various aspects of the use of GIS in a library as quite a good number of papers relating to utilization of GIS in library have been published. The articles entitled, "Use of GIS in Library", really speaking were misleading, however, they have served limited purposes, for the topic of research in hand because most of the articles which have appeared under the banner "Use of GIS in Library" have given emphasis only on following two aspects :

1. Use of GIS in handling cartographic information.
2. Use of GIS in providing information services using census data.

The most of these articles may have been successful in inviting attention of librarians and the user community to the future role of librarians in using GIS technology, not only in training themselves but training the users also. It will also facilitate in giving information services based on spatially referenced data and as per requirements of the users.

However, in the entire exercise and attempts of tracing literature so far, there was not a single write up found about any experiment or discussion on the following aspects :

1. Mapping of Library Collection using GIS.

2. Use of GIS in library networks.
3. Use of GIS in other areas like : (a) Mapping of Geographical facts of class numbers of colour and other classification schemes, (b) Mapping book suppliers and distributors, (c) in user education, and (d) in user study.

Grimshaw,[45] has only touched upon one of the aspects/facts of the esearch project. However, he fails to give any supporting evidence to his concept. The attempt to get more clarifications from him through ersonal contact have not yielded any results. It is possible, therefore, hat Grimshaw is probably quoting the current research topic. There s, thus, every reason to believe that the research topic in hand is a ew area.

NOTES

1. David J. Maguire, Michael F. Goodchild and David W. Rhind, (eds.), *Geographical Information Systems : Principles and Applications*, 2 Vols. (Harlow : Longman Scientific and Technical; 1991).
2. Ian Braken *et al.*, *A Classification of Geographic Information System Literature and Applications* (Wales and South West Regional Laboratory : University of Wales, College of Cardiff; 1989).
3. *International Journal of Geographic Information System (BM),* 1987 - Coppock, J.T. & Anderson, Kerie (eds.) London : Taylor and Fransis (ISSN 0269 3798).
4. Bookscope (CD-ROM) (Woodbridge : Primary Source Media, 1996).
5. Clifford A. Shaffer, "Data representations for Geographic Information Systems", *Annual Review of Information Science and Technology, 27 (New Jersey : Learned Information* Inc., 1992): 135-172.
6. Karl Longstreth, GIS Collection Development: Staffing and Training", *Journal of Academic Librarianship*, 21, No. 4 (July 1995) : 267-274.
7. *Ibid.*
8. Maguire, Goodchild and Rhind, (eds.), *op. cit.*
9. *International GIS World Source Book* (Annual) (Ft. Collins, Co : GIS World Inc., 1989).
10. Longstreth, *op. cit.*, p. 269.
11. P.A. Burrough, *Principles of Geographic Information Systems for Land Resource Assessment*, (Oxford : Oxford University Press; 1986).
12. Longstreth, "GIS : Collection Development", 267.
13. *Ibid.*, 269.
14. Lisa T. Abbott and Carolyn D. Argentati, "GIS " A New component of public services", *Journal of Academic Librarianship,* 21, No. 4 (July 1995): 254.

15. David J. Grimshaw, "Geographic Information Systems in Business", In *Encyclopedia of Library and Information Science*, Vol. 54, (ed.), by L. N. Kent. (New York : Marcel Dekker, 1994) : 187-202.
16. Shaffer, *op. cit.*
17. *Information Technology and Libraries,* Special issues on GIS entitled "Making GIS as Part of Library Services," 14, No. 2 (June 1995).
18. *Journal of Academic Librarianship*, Special issues of Geographical Information Systems (GISs) and Academic Libraries 21, No. 4 (July 1995).
19. Patrick McGlamery and Medisa Lamont, "New Opportunities and Challenges, Geographic Information Systems in Libraries", *Database* (December 1994) : 35-44.
20. Gary W. North, "Will Your library be the spatial data information centre of the future", *Inspel* 23, No. 2 (1989) : 130-136.
21. Patrick Yott, "Mapping 1990 census data", *Information Technology and Libraries,* 13, No. 1 (March 1994) . 63-70.
22. Carl Franklin, "An introduction to Geographical Information Systems : Linking Maps to databases, *Database*, (April 1992) : 12-15 and 17-21.
23. Prudence S. Adler, "Special Issue of Geographical Information Systems, (GISs) and Academic Libraries : An introduction", *Journal of Academic Librarianship* 21, No. 4, (July 1995) : 233-235.
24. Allen B. Cox, "An overview to Geographical Information Systems", *Journal of Academic Librarianship* 2, No. 4 (July 1995) : 237-249.
25. Abbott and Argentati, *op. cit.*, p. 251-256.
26. Li Hunt and Mark Joselyn, "Maximising Accessibility to spatially Referenced Digital Data", *Journal of Academic Librarianship,* 21, No. 4 (July 1995) : 257-265.
27. Longstreath, *op. cit.*, p. 267-274.
28. David A. Cobb, "Developing GIS Relationship", *Journal of Academic Librarianship* 21, No. 4 (July 1995) : 275-277.
29. Theresa C. Strasser, "Desktop GIS in libraries, Technology and Costs: A view from New York state", *Journal of Academic Librarianship,* 21, No. 4 (July 1995): 278-282.
30. Brent Allison, "University of Minnesota : Remote Access and the internet", *Journal of Academic Librarianship,* 21, No. 2 (July 1995) : 283-287.
31. Joseph A. Boisse and Mary Larsgaard, "GIS in Academic Libraries : A managerial perspective", *Journal of Academic Librarianship,* 21, No. 4 (July 1995): 288-291.
32. Joan F. Cheverie "Getting Started Ready Setget Organised", *Journal of Academic Librarianship,* 21, No. 4 (July 1995) : 292-296.
33. Laura Lang, "Mapping the future of map librarianship", *American Librarian,* 28, No. 10 (November 1992) : 880-883.
34. McGramery and Lamont, *op. cit.*
35. Allison, *op. cit.,* pp. 283-287.
36. Pohchin Lai and Charles F. Gillies. "Impact of Geographic Information

Systems on Spatial Data Libraries", *International Journal of Geographic Information Systems,* 5 No. 2 (1991) : 241-251.

37. William P. Kane, "A Hyper card call number directory: Using Stacks to find stacks", *College and Research Library News,* 50, No. 7 (July/August 1989) : 576-577.
38. Irving E. Stephens, "Getting more out of call number . Displaying holdings, locations, and circulation status", *Cataloguing and Classification Quarterly* 13, No. 3/4 (1991) : 99.
39. I.E. Murr, J. B. Williams and R.E. Miller "Mapping Information Delivery Networks", *Library Hightech,* 4, No. 13 (Spring) (1982) : 1-63.
40. Grimshaw, *op. cit.*, p. 198.
41. Jennifer Rowley, "Nature and Applications of Geographic Information Systems", *ASLIB Managing Information,* 14 (April 1994) : 1-4.
42. Mukund Rao, V. Jayarama and M.F. Chandrashekhar, *Organising Spatial Information Systems around a GIS Core* (Database Organisation and integrated analysis using GIS), Special Publication No. ISRO-NNRMS-SP-70-94, (Bangalore : Indian Space Research Organisation, December 1994).
43. Alan Poulter, "Towards a virtual reality library", *ASLIB Proceedings* 45, No. 1 (January 1993) : 11-17.
44. Alan Poulter, "Building a browsable virtual reality library "*ASLIB Proceedings*", 46, No. 6 (June 1994) : 151-155.
45. Grimshaw, *op. cit.*, p. 198.

3

Use of GIS in Library

IMPACT OF GIS ON LIBRARY AND INFORMATION MANAGEMENT

The impact of Geographical Information Systems (GIS) on library is perceived in several ways. The first impact has already been realized in the form of the proliferation of literature which is termed the geographically referenced literature. As estimated by OGRIP Advisory Committee of the Administrative Services of the State of Ohio[1] as much as 80 per cent of all information held by business and the government may be geographically referenced. It refers to information such as census data, mailing addresses, facility layout, telephone numbers, land-use information and details of area networks such as water, electric, gas, sewerage, cable, transportation and telephone systems. Geographically referenced information literally surrounds us. It is observed that the contribution of information on GIS is growing rapidly. As stated by Adler,[2] "Geographical Information System(s) or GIS has in a relatively short amount of time, gone mainstream. GIS is now utilized by an ever expanding array of public and private sector businesses and communities including those in real estate, insurance, utilities, state, local and federal agencies, consumer groups, health care, banking and more".

Handling such a vast area of information is a major task for the libraries as GIS finds unlimited applications in many areas other than traditional ones like natural resources planning and environmental analysis. A good amount of literature is published also in non-geographic information sources.

GIS is being used and required by a variety of users. One would, therefore, expect variety of queries at the reference desks of academic, public or special libraries. This makes it essential for libraries to familiarise with GIS and to use it for acquiring, organizing and disseminating information on GIS.

Librarians and information scientists have already realized this need and have begun to gear up their efforts in anticipation by initiating necessary steps in controlling information on GIS. They have gone further in understanding GIS and utilizing GIS in providing information services on spatial data. These efforts have been reflected in the publication of several articles on GIS and its applications in the journals of library and information science. Several special issues on GIS have been brought out. For example, *The Journal of Academic Librarianship*, has recently published a special issue on GIS (July 1995) entitled *Geographical Information Systems (GISs) and Academic Libraries,*[3] which clearly indicates preparedness of the library profession to face the challenges posed by GIS.

Not only is published information on GIS growing rapidly, but GIS system manufacturing industries are also expanding rapidly. A recent survey carried by Dataquest has indicated that the percentage growth of the market of worldwide GIS systems and softwares revenue for business and sales were 2 per cent of the market in 1990. It is forecasted that the growth would capture 14.6 per cent of the market in 1997.[4] It may not be surprising if users expect to see the actual GIS software in the library in order to train themselves by availing of library assistance and use GIS in analysing information.

GIS has been traditionally used by urban planners and environmental scientists for quite some time. Use of GIS in library in providing information services, is comparatively a new phenomenon. However, with the availability of databases having geographic reference and the need for using them for information analysis it will be found most convenient if GIS is used in the library.

In USA nearly 1400 libraries receive information from the U.S. Government though the depository library programme, which in turn is supposed to make it available for public access.[5] Such databases come in very convenient forms like CD-ROMs and Diskettes. The spatially referenced databases are also available on internet/www which are easily accessible to the users.

Since the users are expecting to get such information in libraries and librarians themselves also have to use it in providing information services, responsible librarians now attempt to fulfil the expectations of being GIS literate, and use GIS in providing information and at the same time acting as resource persons in training users in using GIS.

The impact of GIS on library and information management is manifold. GIS technology is very soon likely to be available in the library leaving no choice for the librarians but to think how to best utilize it for the users benefit. The topic of the use of GIS has, therefore, become more relevant to library management. It addresses many challenges for the present library system and at the same time also offers new opportunities.

USE OF GIS IN LIBRARY

As defined by Aronoff[6], "A GIS is designed for the collection, storage and analysis of objects and phenomenon where geographic location is an important characteristic". GIS is the special kind of information system used for storing, accessing and manipulation of geographic information. The application of GIS having a geographic characteristic can be used in variety of ways in the library.

The applications and use of GIS in library as envisaged by the researcher are as follows :

(a) Library needs to handle a large amount of data having geographic features associated with location of reading material in the library. Traditionally, the library classification scheme is designed to take care of the arrangement of material on the shelves subject-wise to facilitate browsing. However, while actually locating the material or finding out exact location of material on the shelf, there was no user friendly mechanism or system available so far. The GIS finds its application in making it possible by using GIS in mapping the library collection.

(b) GIS can play a very useful role in planning and operating library networks by mapping information resources of libraries which are geographically away from each other but have contentwise relation among them. As an offshoot of this, the use of GIS in thematic mapping of libraries can be successfully used in enhancing access and resource sharing among the libraries.

(c) GIS can be utilised in mapping subject contents against fundamental category of space of Ranganathan's. Colon classification and geographic division of other classification schemes with a view to exploit the geographical features associated with books having similarity in their contents.

(d) Due to its special characteristics and functions, it is possible to apply GIS for the following aspects of library services and functions :

(i) Use of GIS in user study,

(ii) Use of GIS in mapping and networking of book suppliers and distributors, and

(iii) Use of GIS in user education.

(e) Libraries use various technologies in their working and it has been possible to render many services which are based on these new technologies such as OPAC, CD-ROM based services, microform based services and so on. In the same way GIS can be used for giving information services based on spatially referenced databases. In the days to come, libraries are going to receive a large amount of data in electronic form like CD-ROMs, and diskettes, etc., having spatial reference. The library users are also going to seek help from the librarians in getting information by using these resources. Libraries will, therefore, have to think how to design and deliver these services for the readers' benefit. Familiarity with GIS technology will be useful in these endeavour.

It is planned to organise the discussion on the above issues in the following manner:

(a) Use of GIS in mapping library collection,

(b) Use of GIS in providing information services based in spatially referenced databases, and

(c) Use of GIS in mapping/networking the library resources whereas, it is proposed to discuss in short the following uses of GIS in library such as :

(i) Use of GIS in mapping subject contents against geographic features,

(ii) Use of GIS in user study,

(iii) Use of GIS in mapping/networking of book suppliers and distributors, and

(iv) Use of GIS in User Education.

Use of GIS in Mapping Library Collection

The use of GIS in mapping library collection is one of the non-traditional but useful applications of GIS. In addition to several uses, it promises a very useful inventory management tool in the library. GIS, being a powerful database management system can make its benefits available to the library management in terms of its ability to integrate different databases into one environment and also handle information referenced by its location in space. These factors have so many facts relevant to library management and services that the discussion on the use of GIS in the library environment will continue for quite some time.

An attempt towards a similar issue was made by Kane[7] in designing "Hyper card call number directory" at Wayne State University Library where digitized floor maps were used to enable users to find out the location of material in the library by interacting with the Hyper Card Library System. As this is one of the core subjects in the present study it is discussed in full detail separately in the next chapter.

Use of GIS in providing Information Services based on Spatially Referenced Databases

Relevance of Spatially Referenced Data in Library

Libraries have been looked upon as the centres for storage, preservation and dissemination of information irrespective of their being national, public or private. The information that is collected and assembled is all the time for others to see, read and study. This utility has been available since ancient times, when physical forms of recorded knowledge ranging from clay tablets, papyrus, parchment, and paper to microform, digital data and laser disc etc., are being developed and it is to be continued further.[8] As North,[9] rightly puts it "because the library has been the centre for published information through time and is the one place where all the disciplines can truly live together, people will except to find spatial data there also".

What is Spatial Data?

Spatial data could be defined briefly as any information or activity backed up with geographic features. Spatial data is the information source where features or activities are described in terms of their exact geographic location on earth. The data comprises of points, lines, or area features, which are usually heterogeneous. Such data can describe geology, soil, water, weather, forest cover, land-use, pollution and demographic or population statistics. They could be of different scales, dates, areas and other features but it is possible to relate them each other due to their similar topological structure.[10]

Spatial data or spatially referenced information comes from maps, photographs, geocoded databases or from remotely sensed signals received through geostationary satellites. Maps, atlases, globes, charts and geographic tables have been part of library collection since long time and librarians have also tried to provide information by using these sources. However, creating information sources using maps and atlases in the light of user needs and providing them with a spatial context is a voluminous and cumbersome job.

Spatial data can represent text, statistics and geographic information. Geographically referenced features can, therefore, be easily integrated with independent variables like size, colour, accuracy, extent or cost.

Spatial data comes from surveying and mapping organisations governed by state or private agencies. With computers having affordable data storing and handling capacities, it has become simpler to create spatial data generated from maps, photographs and remotely sensed data in electronic form which libraries would otherwise find difficult to create manually and provide information services.

Spatial information is now invariably stored on computers to create digitized spatial databases. In order to use these data special kind of information systems called Geographical Information Systems (GISs) are being developed and used to collect, inventory, manage, analize and to display the spatially referenced data to the users[11]. GIS is computer hardware and software systems used to handle spatial data available on main frame (ARC/INFO) as well as on its front end PC based system (ARC/VIEW)[12]. Most GIS systems convenient for libraries run on 486 and pentium based machines and work in powerful Unix or even DOS based workstations. With a little basic training librarians can attain sufficient expertise to answer reference questions relating to spatial data using GIS.

Producing cartographic map data in electronic form is a new trend in publishing. Public and private organisations are releasing such data on CD-ROM or making it available online or on internet. Maps which are available in electronic form (*Electronic Atlas*) are more conveniently used by the readers using software. They are not only viewed and queried but are also used to create need based thematic maps and download those on floppies for use at home. In USA, the Bureau of Census provides 1990 census data called "Tiger files" to all the depository libraries along with GIS software for internal use and to provide information services to the public at large.[13]

Several associations are evaluating performance and cost of commercially available GIS products. In 1991 a training programme to train librarians in the use of Tiger files called "Tiger test" was initiated by the US Bureau of Census and Government Printing Office by distributing software programmes to libraries.[14] In spring 1992, an ambitious project called "ARL/GIS Literacy Project" to equip librarians with the skills needed to provide access to spatially referenced data was undertaken by Association of Research Libraries (ARL), for its 31 member libraries. The overall goals of the project were to continue to improve free public access to data like Tiger files, educate the public on how GIS technology presents new opportunities for utilizing government information and to develop GIS data and expertise in the librarian community.[15] As mentioned by Adler[16] the goals of ARL/GIS literacy projects were framed to meet the current needs of libraries and users during the time of transition as they are set to initiate network bound services. As referred by him the goals of ARL/GIS Literacy Project include the following points :

- "Introduction of GIS to a variety of libraries (e.g. public, state-based, academic and university libraries in public and private institutions) to address diverse user information needs;
- Development of a team of GIS professionals in the research library community willing to lend time and a expertise to applications, user training and education programs related to GIS;
- Encouragement of connections among federal, state and local GIS users and information;
- Promotion of research, education and the public right to know through improved access to government information;
- Initiation of library projects to explore new applications of spatially referenced data and evaluation of the introduction of these services in research libraries; and
- Implementation of programs to allow institutions that have invested in networking capabilities to leverage the sharing of resources via networks".

In India it will take some more time to see this type of situation in research and university libraries. However, Census of India is in a position to supply 1991 census data in a machine readable form to those libraries which have facilities to use it. Whereas, use of GIS (main frame based) systems for planning and developmental activities has already begun in many government organisations.

It is sure that future database collections of individual libraries will have more and more spatial databases and the library will have to be active in using them and in having expertise in using GIS technology in library. The users on the other hand will also be assertive in putting demands on libraries to answers several queries relating to spatial data through their reference sections. Examples of commonly used questions may be as follows, including some referred to by North[17] :

(i) Do you have a map of the area in which my house is located ?
(ii) What is the pollution status of the area I am living?
(iii) Where can I find fresh water supplies in future ?
(iv) Where are the pipelines located ?
(v) What would happen to the transportation system in my city if there was an earthquake ?
(vi) Where can I open a departmental store in the suburbs of Mumbai where there is a large middle class population?
(vii) I want to begin an earthworm based bio-fertilizer project, where can I get enough hotel waste material ?
(viii) How will water availability be increased in drought areas in India if north-south and east-west rivers are interlinked?

The other significance of having spatial databases in the library, librarians becoming experts and resource persons and libraries providing information services based on these, lies in the responsibility and mission of the libraries and their place in the society. As argued by Cox[18], "Libraries have a primary mission of information dissemination. GIS is a tool that is ideally suited to contribute to that mission". Secondly, there is confidence in users' mind that the library is the first, and logical place to get such information, as libraries believe in freedom of information and information sharing and because of their neutral and trustworthy nature while serving their patrons.

Present Status

Having realized their mission and responsibilities, since early days, libraries housed maps, atlas, charts, globes and the paper bound information sources containing statistical, textual, and geographic details in the form of books, journal and multiple volumes along with other collections; whereas map libraries exclusively stored all the relevant sources of information concerned with geographic reference. They also prepared maps, atlases and charts to meet the information needs of the users.

Traditional resources such as maps and statistical publication used in providing geographic information, were static and there was no option available to modify them according to user needs. The requests put to the reference or map librarians were like "Do you have map of a city?", "Do you have political map of the state or country?" More serious questions were like, "Do you have map of a town drawn per exact scale so that I want to measure area of my plot once again to check if there is any encroachment?"

If the more complicated queries like a map of religious places belonging to denominations like Hindu, Muslims and Christians around the country is asked for. The answer to such questions would be that, there is no such map in the collection and that there are no plans to prepare it. Instead, user would be given a heap of concerned books to get the answer of the question or perhaps he would be guided by the concerned sources to get the details.

As far as demographic or statistical data is concerned series of volumes have been available for consultation but there are no corresponding maps available to support the data as well as its analysis. As such the sources of information cover very broad aspects and the librarian or researcher was not capable of merging, modifying or creating maps using available resources. The situation remained unchanged until the libraries received the geographic related information in printed form.

User Needs and Role of Libraries

The availability of affordable and powerful micro-computer and software in the libraries has made it comparatively simpler for librarians to have computing systems in the library. Soon after the spatial digital data sources were also added to the collection. GIS technology provided the researcher and the librarian a specialised computing platform. Now GIS technology, which is available in inexpensive PC based systems is within the technical and financial reach of many libraries.[19] There are a variety of GIS systems and mapping softwares available so that one can choose a system depending on need and extent of complexity of problem. GIS based spatial data information services make displaying information in map form possible, facilitates analysis and helps in decision-making.

The librarian having GIS technology and the spatial digital data at his disposal is now in a better position to handle even complex type of queries relating to geographic data. He can now answer all the questions referred to earlier under section "Relevance of Spatially Referenced Data in Library".

Now readers are to a great extent familiar with the technology. They come to the library with problems, use appropriate spatial digital data base; create relevant maps as per their needs; plot it with a plotter or download it in magnetic storage form, which can be modified and incorporated in documents.

The library's role would be to acquire GIS technology and software. It also needs to acquire spatially related databases. The library would expect more complex questions relating to them than those that were asked before. In addition libraries need to provide technological platform to their users and at the same time libraries need to acquire sufficient expertise to answer the queries and guide users in getting the information.

Now, computers are freely used in almost all commercial and research activities. They are used in planning, analysis, decision making, monitoring and assessing activities. A large amount of information which has geographical reference could be effectively used if it is mapped and is made available for the readers to handle. Individual researchers or firms need not acquire all necessary spatial digital database and the GIS technology. They should naturally depend on libraries in getting such information handled by making use of GIS technology and analysing the data in the way it is needed. The may also consult the librarian when required.

There is a new trend of publishing data having geographical reference in a machine readable form and which is compatible to GIS. In USA, such data is pouring into libraries. Since 1990 Federal depository libraries have been receiving various kinds of spatially referenced data regularly which can be used with different GIS packages. Such data include Topologically Integrated Geographical Encoding Referencing System TIGER/Line and 1990 census data from Bureau of Census, Digital Line graph files from US Geological survey, economic and agricultural data from government agencies.[20] In addition there are several public domain databases available through internet. Therefore, there will not be dearth of databases required for such services. Libraries only need to be familiar with them and provide guidance to users so that GIS based library services using these databases could be planned and implemented.

Unlike other areas it is not that only after having sufficient needs, can the library initiate services. Library may be only the field where services are planned and started in anticipation and readers respond appropriately or rather overwhelmingly. Reader's needs keeps on changing with time. With availability of information technology the quantum and quality of needs have not only increased but have also become more complex. In case of spatial databases and GIS technology just availability itself has created a large needs of spatially reference data which otherwise was not realised in the absence of this technology.

The role of the library intending to start GIS based library services would be consolidated as below :

- Assess the users needs and level of use.
- Acquire suitable GIS systems and a software.
- Attain sufficient level of literacy in GIS and acquire proficiency in using GIS technology by the librarians and library staff.
- Provide GIS technology platform for the readers.
- Initiate services based on spatial data to meet the common man's expected queries.
- Conduct awareness programmes for users on potentialities of the services.
- Acquire spatially referenced databases depending on the requirement of the users and provide the access to users.
- Provide access to users to www on internet, in order to utilize database available on-line from them.

While discussing the role of the library involved in GIS, Abbott and Argentati[21] have stressed the need to evaluate user needs and the level of use of the client community and to device a "Service Mix" which best addresses those needs with available resources. They assign the following three major roles for the library in the endeavour :

1. "Developing Collection of GIS related materials and spatial information,

2. Providing access to spatial information software tools, and
3. Facilitating intellectual access to all concepts and applications."

Challenges and Opportunities before the Libraries in providing Information Services using Spatially Referenced Databases

Voluminous literature on GIS or GIS technology itself, is available. It is very much in the library leaving no option for libraries but to accept it. The question is how libraries should optimally utilize GIS. If it decides to accept GIS as one of the tools in providing information relating to geographic details then it is likely to face some challenges. The major challenges before the library, as this study emphasises is to provide services with high degree of accuracy and also with authority. Other challenges, including those discussed by Cox,[22] are as follows :

(i) How to obtain spatially referenced data bases relevant to user needs and how to establish accessibility to such data if available on-line or through internet/www.

(ii) How to achieve sufficient literature and expertise in understanding and using GIS technology. Most library professionals show their preparedness to learn new things and acquire new skills and to motivate the few who show resistance to change.

(iii) Identifying and choosing the right kind of GIS and software which would be competent enough in handling a variety of queries ranging simple to complex ones.

(iv) Copyright problems as databases are used to create or plot maps depending on user needs and as the downloaded results or output is likely to be used for commercial use. The library has to look into all legal aspects of copyright of the database manufacturers before permitting users to copy them.

(v) How to assure quality and accuracy of map or outputs created by patrons and library staff as even small errors may create misunderstandings and chaotic situations.

(vi) How to mobilize financial resources to buy GIS software, and the latest hardware and to acquire commercially prepared databases.

(vii) Creating spatially related database for the library wherever no data is available.

To overcome all the above challenges and to organise "GIS Cell" in the library to provide services, the library needs lot of efforts and skills. Libraries have shown their competence on several occasions when new technologies were adopted in the libraries and it is expected that they will overcome the barrier this time too. Answering simple queries like viewing map and atlases will not be difficult at all with just an introductory knowledge of GIS. However, to use full featured GIS it is necessary to have acquaintance with high level operating systems like UNIX. For creating new databases a segment of the library staff need to have advanced education in computer and geography/ cartography.

As stated by Cox, GIS has forced us to adapt map and geographic data in new and innovative ways. Using maps in the electronic form and by rearranging data in a variety of forms, one can create infinite number of geographic map depictions. GIS platform in the library opens many new gateways and provides several opportunities to the libraries for especially contributing their share in planning and decision making in the area of handling geographic information, which they did not avail earlier. It is possible to answer a variety of queries put by patrons working in different fields. The library can take up a number of projects relating to mapping and analysis, using GIS and databases, can offer consultancy in GIS technology and guide users in facilitating outputs as desired. GIS provides librarians new opportunities in devicing a new library inventory management tool as is advocated in the present study. The library, with its continued research can promote newer GIS applications in other fields as well. In a special library which is working exclusively in geographic area, the Librarian can have a number of responsibilities and opportunities while working with GIS. However, other libraries irrespective of their type Academic, Public or Special will have at least one unit to use GIS and offer services to their users based on spatially referenced databases. Thus these libraries will have ample opportunities in dealing with GIS. The opportunities, as discussed by Cox are referred as follows[23] :

(i) Libraries can function as information distributors by providing continued access to data or to serve as depository centres,

(ii) Libraries can become information brokers,

(iii) Libraries can serve as GIS project facilitators,

(iv) Libraries can collaborate with other agencies in providing information,

(v) Libraries can develop need based databases drawing on large databases prepared by the government and other agencies which will be cost effective,

(vi) Libraries can offer suggestions in removing errors found in databases prepared by government and othér agencies and in bringing more accuracy and quality,

(vii) Libraries can become GIS service providers, and

(viii) It can undertake research in using multimedia along with GIS to present data more dynamically including utilizing animation techniques for the same.[24]

Since the last 15 years, GIS is diffusing into all departments especially in government departments as a powerful inventory management tool and decision support system at all levels. GIS has proliferated from its main frame version to even desktop level in government agencies and private business which use it also as decision support system. It is in fact enfluencing into the daily life of average citizen as noted by Cowen.[25]

Library Services using Spatially Referenced Data and GIS : Future Prospects and their Managerial Perspective

The organisation of GIS based spatial data based service would depend on three factors. (a) Type and range of GIS software available, (b) Range of spatial databases available with the library, and (c) Level of patron using the services. Among these, the third factor is an important one, as with any change in the patron use level the other two have to be changed. Therefore, the library's first and foremost task is to assess the user needs for this kind of services. The user needs could be categorised into three levels of services. (1) Introductory, (2) Serious, (3) Complex. The library ought to take a policy decision about where to stop or whether it should provide full featured GIS services to its clientele. The three levels are discussed as follows :

Level 1 : This is introductory level where users are exposed to user friendly mapping software to practice the basic mapping function of the software. Here, the user can explore and display spatial information. The system typically contains value added geographic and attribute data which can be used to get an idea how to create thematic maps. Examples of typical GIS software used here are Map Expert and Map 'n' fact.

Level 2 : Here, the user handle problems having more complexity. In this level the system chosen provides more sophisticated functions, such as manipulation, querying, customising and displaying value added geographic as well as attribute data. To handle these types of querying librarians need to provide some assistance to users. This type of service could be similar to services given by US Depository Libraries using Tiger/line files 1992 or census summary tape files (STF). Example of GIS software used for this level of usage are ARC/View and Map info. Training programmes similar to ARL's GIS literacy project would be useful and encouraging for librarians who want to introduce this type of service.

Level 3 : The most complex level of service would include use of full fledged system like ARC/INFO and users require extensive training before using it. Here the system is modified to capture, input, store, manipulate analyse and display value added public domain or unique geographic and attribute data.

The above levels are based on 3 levels viz. initial, intermediate and research as defined by North California State University Library in their service policy statement pertaining to the use of GIS, as referred to by Abbott and Argentatic.[26] Examples of services to be provided under the given 3 levels are as below :

Level 1 :
- (i) Browsing and exploration of digital/ electronic atlas,
- (ii) View and print map of various countries of the world, and
- (iii) Creating a specific map with certain location features, in mind.

Level 2 :
- (i) Population and demographic study and analysis at municipal level spatial databases,
- (ii) Agriculture or soil mapping, and
- (iii) Mapping of pollution threatened areas at state, regional, national or global level.

Level 3 : Library undertakes collaborative GIS projects along with University or private agency on a wide range. They help decision makers to assess sources of environmental pollution and risk contaminants in water supply; to develop transportation routes or develop land development plans for cities having rapid population growth; or to study the disease spreading pattern in a particular area.

Project feasibility requires a series of questions to be answered through GIS mapping and analysing steps. For example, if one wants to establish an export oriented floricultural company he would like to consult the library having relevant information on the topic to seek advise. This problem needs analysis and a series of mapping on the following details :

1. Identify range of land suitable for rose planting based on some defined parameters.
2. Identify area where moderate rains occurs and climatic conditions are also suitable for rose planting.
3. Who are owners of such land and their details ?
4. How water is made available from nearby dams or canals to the place ?
5. Levels of land and wind pattern of the region which need be suitable for erecting a greenhouse structure to control the climatic conditions inside the house to get better flower yields.
6. How the place is approachable by roads ?
7. Availability of insulated transport systems nearby.
8. Which airports are available near the plantation, what is the distance and whether they are internal or international airports ?
9. Availability of manpower in the area.
10. Identify the existence of similar plants nearby.

With the availability of relevant spatial databases and GIS, library systems can handle such projects and provide very useful and exact information which will be directly used in decision making or planning of such projects.

Collection Development

As far as collection development is concerned, the library first needs to concentrate on two aspects. Firstly to acquire sufficient literature on GIS by purchasing handbooks, reference books, introductory books and journals so that librarians and users can educate themselves. Secondly, it needs to acquire commonly used spatially related databases on CD-ROM to be used by users as well as by library staff so that most of the reference queries would be answered.

The databases concerned with demography and census data are

generally prepared by government agencies and are available on machine readable form at nominal costs. There are variety of GIS software available in the market. A large number of data resources are available on Internet/www. ARL Electronic Atlas is a very useful source of database where a group of libraries are collectively developing internet accessible maps and imagery of USA.[27]

The printed sources to get information regarding databases, GIS literature, and GIS softwares include :

(i) Geo Info Systems (Monthly Journal),
(ii) GIS World (Monthly Journal), and
(iii) GIS World Source Book (Annual): A comprehensive reference source virtually covers all aspects of GIS.[28]

The present study has presented a detailed review of literature on and related to GIS in the chapter "Review of Literature". The sources available on the internet are as follows :

(i) Geoweb (MAJORDOMO@CENSUS.GOV) focuses on efforts to build a comprehensive set of GIS data,
(ii) GIS-L (LISTSERV@UBVM.CC.BUFFALO.EDU) discusses all aspects of GIS such as data and hardware, software and technical issues, and
(iii) Maps-L (LISTSERV@UGA.CC.UGA.EDU) covers mapping softwares, geographic data availability, reference queries and cartographic information issues.[29]

Staff Training

Development of adequate and suitable staff is the crucial factor in organizing GIS related library services. With the availability of spatial data sources from government agencies readers would naturally like to take advantage by framing related questions. The reference staff themselves should be well equipped with necessary hardware, software and database and ready with the expertise and skills to answer them. With this need in mind, the Association of Research Library organised ARL/GIS Literary Project to train librarians and make them knowledgeable in using the databases and making them accessible to the users with necessary guidance. There is a warehouse of literature on GIS instruction. These instructions are meant for graduate and postgraduate level. The National Centre for Geographical Information and Analysis (NCGIA) has played an important role in developing curricula for classroom use.[30] In USA many Universities and Colleges offer GIS instructions. ARL/GIS training programme was specifically designed to train librarians and library staff. ARL/GIS Literary Project was sponsored by ARL and Environmental Systems Research Institute (ESRI). ESRI provided software, database and training for librarians. In India considerable efforts are on in providing training in using GIS in various areas. These efforts, at the moment, are concerned only with the use of GIS technology in handling spatial data. However, there are indications of introducing formal and comprehensive GIS educational programmes at the potential institutes like IITs, System Research Institute etc. and it is hoped that they will succeed in it.

The main objective of staff training is to make them resource persons in providing information and also to guide users in getting the information as emphasised by Longstreth[31] "Staff must know more than how to operate the GIS software, they need instructions in issues of GIS theory, GIS databases and GIS applications in a discipline".

Hardware and Software Requirements

Having appropriate and efficient hardware is another important factor in implementing GIS based services in the library. The software and the hardware requirements would be guided by the level of GIS related services the library wants to introduce. Hardware and the software requirement to establish a GIS based library service work station considered at University of California at Santa Barbara[32] are as follows the typical requirements are for entrance level and it is characterised as PC based GIS service. The level requires the hardware as follows :

(i) At least one image processing programme, and GIS package.
   (a) For PC based image processing software XV (an excellent but inexpensive software one which is based on UNIX operating systems.
   (b) IMDISP and IDRISI are good inexpensive choices for DOS environment.
   (c) ERDAS/Imagine is another choice for DOS version, but a little expensive.

(ii) GIS software such as ARC/INFO (a product of ESRI) or its front end product ARC/VIEW; Atlas/GIS from strategic mapping, Map-Info and Sammamish are worth considering.

(iii) Hardware: the typical hardware requirement recommended for a library participating in US depository programme of Government Printing Office who receive TIGER files at the level of technology prevalent are as follows :

- 486 DX CPU IBM Compatible (Maximum),
- 50-66 MHz. (It is an important measure of micro-processor performance),
- Floppy diskette drives both 3.5" and 5.25" high density,
- 200 M, SCSI Controller Math Co-processor graphic card,
- Minimum of three free expansion but broad slots. One or more additional hard drive bays desirable. 2 serial port and one parallel port, ZIP, CPU upgrade socket,
- 16M of RAM,
- No less than 1M display memory,
- MSDOS 6,
- Windows 3.1,
- CD-ROM - Compatible with IS09660, Multimedia supporting standards are 300 k/byte per sec. transfer rate double speed support CD-ROM XA support, 16 bit sound board for PC, Adlib or sound blaster,
- Mouse to support Microsoft windows,
- Black and white laser printer, Colour printer or plotter,
- Connection to Internet, optical fibre/ethernet or broad band, NIC (Network Interface Card) Modem (9600 bps meeting V.32, V. 42, V.42 bis or MNP 5 standard compatible with Hayes "At" command set,
- Large screen (15" minimum) high resolution, Colour monitor super VGA compatible, and
- Digitizing tablets and scanners.

The configuration referred to above need to be revised in the context of the current level.

The overall cost of having the above items to establish first level kind of GIS based spatial data library services would range from $ 3000 - $ 6000 (i.e. between Rs. 1 - 2 lakh).[33]

The second and higher level of service station requires UNIX operating system, a higher level image processing software, ARC/INFO GIS software and public domain software GRASS (Geographic

Resources Analysis Support Systems). At this level work station will require minicomputer from SUN, DEC or IBM 600. And the price range for the work station would range from $6000 to $7000.[34]

The idea in giving the configuration is to get an estimate of how much a work station will cost the library if it decides to initiate GIS based library services using spatially referenced databases.

The cost of only GIS software (ESRI's widely distributed flagships GIS product) such as the PC based software ARC/View 2.0 in a special package of data for publicly accessible libraries is $ 495. Atlas map making for windows (Strategic Mapping) costs $ 495, Map-Info for Windows (Map Info Corp.) costs $995 and so on.[35]

Use of GIS in Mapping/Networking Library Resources

An ambitious collaborative project launched from Minnesota's John R. Borchert Map Library, Automated Cartographic Information Centre (ACIC)[36] provides access to digital spatial information which can be remotely accessed. It is also possible to manipulate this information. The centre is also capable of handling the material found on internet or World Wide Web and of providing remote access to users. The Unit of Minnesota library actively participated in ARL/GIS Literacy Project. Nearly 30 member libraries participated in the project and jointly developed ARL Electronic Atlas. The project's aim was to systematically prepare maps depicting demographic, environmental, trade and other data and make these maps and other data available for viewing and downloading off the internet. The initial phase under development include providing maps of census data for major urban areas in the United States. The prototype is accessible on Worldwide Web. Their home page displays an index map of U.S. [Fig. 3.1].

The ARL Electronic Atlas project inspired the researcher to apply the idea in mapping the library resources in a given area or in other words creating a "Virtual Library Network". The issue is discussed with a view only to introduce the idea. It is not a reality so far. Suggestions are presented for the ways it could be implemented.

In this discussion entire library resource has been considered as one entity for the purpose of mapping. It is different from mapping of library collection (Chapter 3), where a book has been considered as one entity for mapping.

A lot of co-ordinating efforts are involved in the process of creating the formal library network. For the participating libraries it becomes

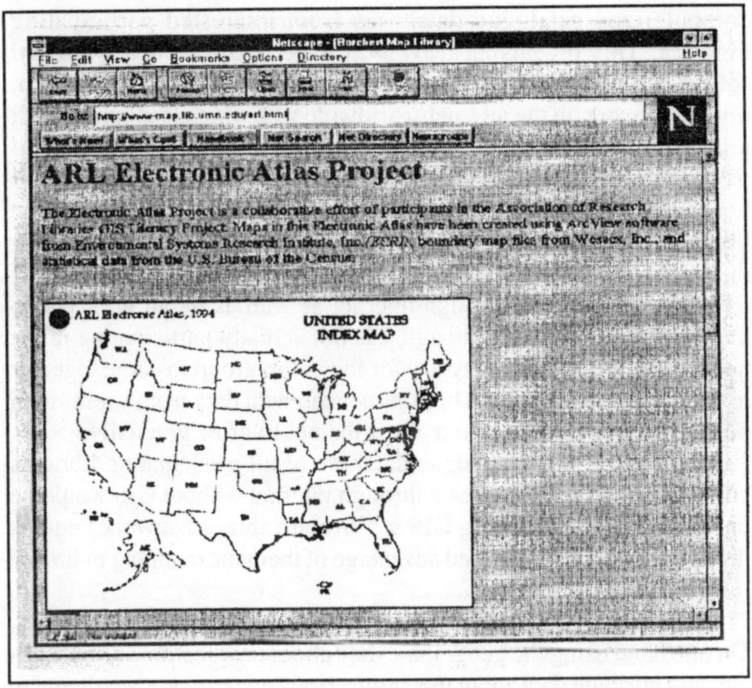

Fig. 3.1 : ARL Electronic Atlas Project. *Courtesy* of Brent Allison : University of Minnesota : Remote Access and the internet, *Journal of Academic Librarianship,* 21, No. 4 (July 1995) : 284.

mandatory to agree to terms and conditions of the network and they are required to share the cost of establishing the network. Whereas at the individual level they have to prepare and upgrade themselves according to standards laid down by the network and are required to have networking and communicating hardware. Thus the formation of local and wide networks have been observed to be extremely difficult. As compared to these difficulties it is felt that it is easier to form "Virtual Library Network" with only the minimum condition that the participating library has computerised bibliographic database and has an agreed minimum communication and computer hardware required to access the internet. It would be a network of databases without participating libraries being connected with each other using costly hardware.

The concept of forming virtual library network of course requires leadership by an agency or library association who will try to get computerized catalogue databases from interested participating libraries. This organising body will have adequate hardware and software plus necessary communication system to put the virtual library network on the internet/worldwide web which is mounted with a little higher level Geographical Information System. The data of participating libraries also ought to have a minimum level GIS based library information system which has incorporated the location of the library on campus of the parent body browsable by the outsider. Having this preparation the network would be ready for being accessed by participating libraries through internet as well as by an library who has internet facility even though it is not actually participating in the network. It is also not necessary for the entire group to come together in forming the net; it could be commenced with data from a minimum number of large libraries and allow the net to grow gradually.

Forming Union Catalogue database of all participating libraries and retrieving the information through internet without GIS would be possible. However, having GIS a "Virtual Library Network" equips the network to provide added advantage of thematic mapping of library resources using GIS.

While retrieving information one can have two major advantages. On one hand using the geographic data about the locations of libraries etc. and attribute data about their collections one can prepare thematic (geographical) map of libraries where the collections on interested subjects are held broadly, one can find the availability of specific required documents among them and decide about obtaining the hard copy conveniently. On the other hand one can go further to browse through collection of the library by using GIS based library system and try to locate the document upto a shelf level. As such it is interesting to use such facility and use the information in retrieving document directly or indirectly.

The support of GIS for the network system makes retrieval more meaningful as it is backed up with geographical features through thematic mapping facility.

Information about the availability of required material alongwith geographical and other mapping information at more than one single locations the system facilitates the user to decide.

(i) Which library is nearer to the user, with distance from the accesser.
(ii) Availability of transport.

(iii) Convenient train or road routes.
(iv) Availability of courier facilities with those libraries.

The above points are very much relevant in retrieving documents from the host library in the minimum amount of time, with minimum efforts and expenses.

If the host library has GIS of a little higher level or a more powerful system, it would also add values to the participating local libraries having minimum level of GIS based library system. This would help in increasing retrieving capacity of same data when accessed through network system or through internet due to its high level database management system, which would be advantageous to local users of the participating library.

Secondly, it would be possible for local library to use the GIS system for converting the traditional system into GIS based library (location accessible) system by providing locational data and floor maps to host system which could digitize the map and provide GIS based library retrieval system to local library through internet.

Hardware and Software Requirements

The centre or host of the network requires the following hardware :[37]

- A 486 DX 2/66 File server 64 MB RAM and 10GB storage space.
- Sufficient number of work stations of 486 DX 2/66 with 32 MB RAM with colour monitor to facilitate networking within the centre.
- Printer.
- Digitizer.
- CD-ROM readers.
- GIS software.
- Ethernet software facilitating connectivity to internet supported by network software.
- Modem.
- Bibliographic data from participating libraries.
- Map data pertaining to area/region forming the WAN/LAN network.
- Locational data of libraries.

The above hardware however needs to be upgraded with prevalent available configuration.

Organisation

The map of a particular region under consideration is digitized. Locations of all participating libraries are also incorporated in the digitized map. The digitized map is linked to attribute data such as details of libraries, their rules plus their catalogue data by using a suitable GIS software which includes the geographical location details of library on campus map. With this, the system could be ready to be accessed by other libraries once the centre puts the network on the internet. The library seeking access will only need internet connectivity and then it can use the home page created by centre on the internet. While accessing information, the library can prepare a map of libraries according to its need and search the material in their catalogues and send the request for document delivery or decide to obtain the documents as per its convenience.

Initially some major libraries can come together with their catalogue data to form the network at one centre. The centre needs to equip itself with the necessary hardware and software to map the data and make it available on internet. The data could be updated regularly if the right is given to participating library to add their latest data to the database already existing. This would enable the centre to build the union catalogue continuously. Subsequent to the formation of the network, other small libraries may be allowed to join after completing minimum formalities. They may be given some incentive on being contributors of data.

Advantages

1. A virtual library network could be established without going into formation of formal network and with minimum investment.
2. Libraries working in a similar area can automatically come together through the thematic mapping facility. It would be possible to form small groups of libraries among the large group depending on specific areas of interest which could be accessed by any library which has access to internet.

3. Downloading, cataloguing and customization of data would be possible for local use and retrospective conversion.
4. The system would make open the world of libraries to the users of local libraries.
5. The local participating library can make use of sophisticated GIS system available with the network host and use it to retrieve their own data with more powerful OPAC features.
6. The system would facilitate resource sharing to a great extent, which would help in conserving resources at group and individual levels. Subscriptions of periodicals also could be monitored in a group of libraries in order to conserve the funds for periodicals and to strengthen the information resources if database of current periodicals is prepared.
7. The system not only facilitate thematic mapping of library resources but allows a browser to browse the location of libraries on campus map and also facilitate, browsing the library and its collection upto the shelf level.

In addition to details of hardware and software discussed in the above referred systems a detailed list of GIS Vendors, their products and other useful information has been appended at the end of Book in Appendix - I.

Use of GIS in Library in Other Activities

Use of GIS in Mapping Subject Contents against Fundamental Space Category of Ranganathans' Colon Classification

Introduction of Idea

In providing classification structure every classification scheme tries to provide geographic treatment to the subject wherever essential. Different ways are used to indicate geographic divisions in class number. In UDC it is provided in the parentheses. Others provide them under either quotes or attach the geographic number which is separated by "Zero" whereas Ranganathan has provided it under "S" facet which is one of the Five Fundamental categories of knowledge classification viz., PMEST in structure of the colon class number. He has applied the geographic feature to each and every subject schedule.

However, the classification system does not provide any mechanism to exploit the data having locational or geographic concept in mind. Perhaps computer would make it possible by sorting the data on the geographical aspects provided in the class numbers. But if GIS is also available to support the computerised catalogue the system would provide unique analytical features with graphics in the OPAC, which otherwise would not be possible.

It is possible to enhance the geographical features of classification scheme by presenting the subject geographically or with digital map. Ranganathan as the pioneer classificationist while accepting the geographic relation with each subject has indirectly indicated that there is potential in showing geographical feature which could be exploited further.

This potential may be optimally utilized by relating Geographical Information System with the geographic treatment of the library classification scheme. At the moment these features are static and users have no choice in using them when they require to present the data with analytical model in geographic context. Geographical information systems makes it possible in making geographic feature dynamic by transforming it into geographical information system environment and adding as a complementary feature to the classification scheme.

Benefits

There are several advantages if GIS is applied methodically :

- It can find new relations between world knowledge and geography.
- Geography is more understood if it is with visual backup it will be possible to map and compare a subject against any parameter, so that the idea and concept could be cleared.
- It can provide automation to the geographic facet of the colon classification and provide benefits of special analytical features to the readers.

Examples : As classification gives geographic treatment to the subject when it is essential, the care is also taken by bringing books on the same subject having same geographic context together using geographic subdivision as a part of class number. This, however, does not highlight coverage or effect of any subject contents of a book, in broader or narrower sense, across the geographical boundaries. By using GIS in relation to the geographic facet it will be possible to emphasize this aspect and demonstrate the special significance of the geographic facet, as was envisaged by the great classificationists like Ranganathan in the treatment of a subject. Some examples based on the above idea are given below :

Example 1 : Suppose the library has books which discuss medicinal plants grown in different parts of the world. The library has this information spread in different books. Although the books have geographic number and although books in the same subject as "Medicinal Plants" are arranged further with geographic numbers and corresponding cards in classified cards are grouped accordingly, it does not give a global view.

Different books have information about the geographical locations and availability of medicinal plants in the respective country or area dealt with in respective books. Such information pertaining to different parts of the world can become comprehensive sources of information available for the research to build an analytical model while handling such information in GIS environment. Even the library staff can initiate the creation of diverse types of sources of spatially referenced database and make it available for the users. The analytical model or such spatial information would not be available in any book.

Example 2 : Today many developed as well as underdeveloped countries are affected by terrorist activities. A large amount of information is generated by different countries in the form of reports with geographical features of terrorist activities in the respective country in the separate documents. It would be possible for a researcher to consolidate this information and handle it with GIS. It would facilitate to have comprehensive data to understand activities of terrorists in different countries across the world and by comparing local and global information it would be possible to undertake appropriate measures to curb their activities.

Example 3 : Library could have information on different books about oil fields or petroleum resources with their geographical features. If this information is consolidated and used through GIS in order to create an analytical model it can give vital information on whether there is any geological continuity among these oil fields, and it can throw more light on exploring fresh sources of oil in different parts of the world or to understand the behaviour of petroleum streams under earth's surface.

There could be plenty of examples of such prospective uses of GIS in exploiting subject contents in terms of geographic treatment by using GIS in catalogue retrieval system. Some of them are listed below :

1. Study of train networks in different countries and exploring train transport among the neighbouring countries.
2. Study of spreading different diseases in different parts of the world.
3. Studying Drug trafficking channel in the world.
4. Mapping draught regions around the world.
5. Mapping Earthquake zones in the world.
6. Mapping Indian localities in different countries or mapping IIT graduates migrating to different countries.
7. Mapping nuclear or other warfares deployed by the different countries.

Implementing the Idea

How may GIS be applied to exploit the space facet in the colon or geographic facets of other classification schemes? Colon has provided comprehensive geographic table in its revised version of the seventh edition. UDC and DC also have revised area tables on the same lines. The geographic numbers could be considered as code or location numbers, which could be used as linking devices between map data and the bibliographic data. The map of world is digitized using suitable GIS systems or electronic map in GIS environment. The geographic numbers given in the area table are matched with location on the digitized map. The use of such digitized map would be made available for the users through OPAC optionally whenever users want such mapping and comparative analysis for their study. The digitized map interacts with bibliographical data or extraction information through GIS software. The results are displayed in graphic form as well as in bibliographical and other geographical details one by one or simultaneously. It could be possible to download the results onto a floppy which may be further used for analysis and research. The system would be available for each and every subject for application.

The idea could be implemented for all classification schemes that provides geographic facets in structure of the classification numbers. Such mapping of books would be possible where geographic numbers are a part of classification number.

There are some more areas where GIS finds its applications in facilitating an altogether different scheme in managing the library. They are :

1. Use of GIS in Users Study,
2. Mapping of Booksellers and Suppliers, and
3. Use of GIS in User Education.

Use of GIS in User Study

The first law of library science of Ranganathan says that "Books are for Use". The library contains many things other than just books. It would be appropriate to say that in a broader sense the "library collection is for use" and the collection must be used as lot of efforts are put in to make the library services appropriate to suit the user's convenience. It is possible that some users prefer not to come to the library if the situation permits them to access the library services from homes or offices. The librarian community would definitely not like this and would like to do something to avoid such an attitude. Attempts are normally made to study the user needs, problems, behaviour and other details to overcome such problems.

Now-a-days, large amounts are spent on purchasing technological products for user benefits. These include computer, CD-ROM, Microfilm, Reprography, Telecommunication, Bar code, Security technology and others with a view to suitably automate or mechanize the library processes.

Most of these investments are one-way process unless the library is keen to earn funds by offering these technology based services to the users at cost. It is unprofitable and unproductive for libraries not to earn money out of their services. The marketing of library services is recommended. Marketing involves attracting users to utilise the library and its services. Several techniques are used to study user's behaviour and needs. It is possible to use GIS also for this purpose. Institutions, academic and research organisations and corporate users are potential customers as far as the library's income avenues is concerned. All these users are geographically spread around the location of the library. It is necessary to identify them and assess their needs. For such user surveys, geography comes into the picture and thus GIS. Using GIS user organisations could be mapped, as well as factors like vicinity, availability of transportation facilities, their needs against availability of resources with the library operation of courier and communication services, etc. The study would facilitate decision making right from identification of prospective users to enhancing library services.

Using the above locational data a digitized map could be prepared and could be linked to the attribute data containing details of users. The system could be used to carry out necessary analysis and the results could be used in decision making regarding planning library services for corporate and other client users located around the library.

Use of GIS in Mapping of Book Suppliers and Distributors

GIS can also be used in co-ordinating book trade or library supplying activity. Many times library faces problems in timely supply of books mainly due to difficulties faced by book suppliers in reaching and supplying books to a library, which is a geography related problem. In metropolitan cities like Mumbai, movement and transporting are acute problems faced by the businessmen and traders. Not only do libraries face difficulties but individual readers also find it difficult to reach book shops for personal purchases. Establishing a book shop on the campus of the academic institution could be an easy solution. However, it is observed that there are advantages and disadvantages in having such arrangement. Such situations demand a scientific study of book supplying activity. If locations of the library and the book suppliers are taken into account then the geographical study could be undertaken. And where geography is one aspect of study GIS provides solutions. Like other business activities it would be possible to have a comprehensive data about the book suppliers along with their geographic locations with which it would be possible to prepare a map of book suppliers and some analysis could be done with different variables like their location and distance from the library, accessibility by road or railways and availability of public transport for carrying men and material to the library. The system would provide some definite guidance useful in decision making for both of them so that a model could be designed to decide which book seller would be most suitable to give services to the library.

GIS thus would make it possible to have a network of the book suppliers so that book selling activity would be coordinated in such a way that the book requirements of a particular region or city could be monitored. The idea of using GIS for networking of library resources where libraries are mapped against geographic feature of the region has already been discussed. It could be, therefore, possible to combine two maps viz., map of library network and the map of booksellers network and have a superimposed map. This would facilitate the coordination of book purchasing activities of libraries participating in the network with book supplying activities of booksellers network to find a match in identifying participating booksellers located near a particular library. With such a system it would be possible to implement ideas like cooperative acquisition, to establish a formal network of libraries and book distributors, and to have electronic mail ordering systems and establishing direct contact with publishers through combined channels of these networks. The GIS could be used to a maximum extent in planning and decision making in these activities.

Use of GIS in User Education

Use of GIS in User Education is another area of application of GIS in library. In planning user education programme one of the main features is to make the readers aware of the locational details of the library premises and the organisation of reading material in the library. In addition to the traditional methods like literature, bulletins, lectures and tour now-a-days more sophisticated techniques like tape slides programmes, video films and Computer Assisted Instructions (CAI) programmes are also used in users education programmes. The visuals used in these methods emphasize the locational aspect and geography of the library. It is, therefore, felt that GIS could be effectively used for this purpose. In this connection the recent articles by Laing and Mander[38] entitled "where history and technology meet" deserves a special mention. The article has described a system where a software named "Hackneys on disk" has been developed which integrates map and visual images for use in record offices, libraries, museums and educational services. The system consists of putting map of Hackneys Archives Department the record office, local study library and Sutten House dating 1535 serving Hackney on window posed application system for the integrated storage and retrieval of period map and historic images allowing the researcher to study the history topics for a given locality. The salient feature of the system is to see the location of the historic point on the area map and also to see the corresponding details with text and images. The software allows a variety of ways in which to retrieve the information based on photographs, dates (being historic value), subjects, keywords and areas.

The package which is commercially available under the brand name, Heri Tage Access 2000, which includes all the following software requirements to establish and maintain visual collection :

- "Creating and maintaining image and map databases, including definition of fields to be stored.
- Designing and changing viewing screens.
- Scanning of maps, photographs, water colours, documents etc.
- Capturing of video images.
- Linking images to maps.
- Calibrating maps.
- Building specific multimedia applications.
- Developing catalogues".[39]

In the same way GIS could be used to prepare visual programmes through computers and GIS system as part of user education programme. The system would involve, features like details of shelving areas, locations of service points, views of the interior of library. In short a software could be developed with the use of GIS which can give the actual but simulated view of the library premises. Users may be allowed to interact with the system to be introduced to the library and its operation through the programmes of different levels like library orientation and use instruction.

It is also possible to incorporate other technologies like multimedia and animation technologies with a view to make programmes more effective and attractive too. The programmes could be extended to higher levels especially in locating material on one subject available in different forms and types located at different places in the library.

GIS can also be used in mapping the library collection which can provide a simulated view of library stacking area in order to facilitate the retrieval of material along with their locations. The study proposes to discuss this aspect in detail in the next chapter.

Conclusion

This chapter has provided a detailed exposition of characteristics, functions, utilities and requirements for GIS to apply in the library through various applications stated above.

It is seen that there are clear indications that libraries would be looked as source of information for the services based on spatially referenced databases. This is because :

(i) Library does contain a lot of information on GIS and library is likely to be recognised as training centre or warehouse where users would like to update their knowledge about GIS and its applications.

(ii) Now-a-days there is a trend to introduce data including geographic or spatial nature on CD-ROMs and disketts by the government and private agencies. The future library collection will have a large number of such databases.

Librarian's role in providing services based on these types of databases is significant and it is a timely need to accept this challenge by them. The libraries, therefore, need to equip themselves with relevant technology, databases and librarians to equip with adequate expertise. There is a lot of potential in this type of use of GIS in library in future. These services are required to be systematically planned and widened. Database collection is also to be strengthened and new databases to be created through the initiatives of the librarians.

As regards use of GIS in mapping/networking of library resources is concerned it is concluded that organising library Network by using computers and telecommunication technologies and providing services has already been possible. However, thematic mapping of libraries using GIS and forming Virtual Library Network on internet brings a lot of convenience and economy. It can solve problems faced in forming networks at initial level and can provide direct advantages of networking without going into actual procedural formalities. The suggested system brings the geographic factor which is very vital one in operations of post-networking situation. It is, therefore, confidently stated that this application of GIS will be widely used by the libraries in forming their network platforms. Library associations or co-operatives can perform a very significant role in establishing the network in the suggested manner.

The application of GIS in mapping subject contents against geographic facet of Colon and other classification schemes, provides an altogether new mechanism in the area of classification and cataloguing for accessing and browsing the library collection and the subjects. This facility is an innovation and is essential for efficient and up-to-date library service. This type of application will have a wider impact on dissemination of knowledge from books and will influence and have impact on library management. Idea of using GIS for exploiting the spatial facet of classification has a lot of potential for further research in this direction. As regards other applications discussed in this chapter such as : Use of GIS : (i) in mapping book sellers and distributors, (ii) in user study, and (iii) in user education programmes, it is stated that these aspects, no doubt could be managed by using computers but by using GIS a very useful "geographic" dimension comes into picture which provides an innovative way in managing above aspects.

NOTES

1. OGRIP. Advisory Committee's First Year Report, Department of Administrative Services, State of Ohio, 1990. quoted in Franklin Carl, "Introduction to Geographical Information Systems : Linking Maps to Database" *Data base,* 15, No. 2 (April 1992) : 12.
2. Prudence S. Adler, "Special issue of Geographical Information System (GIS) and Academic Libraries : An Introduction" *The Journal of Academic Librarian,* 21, No. 4 (July 1995): 223.
3. *The Journal of Academic Librarianship,* 21, No. 4. (July 1995) : 231-338.
4. Adler, *op. cit.*, p. 233.
5. Patrick McGlamery and Melissa Lamont, "New Opportunities and Challenges Geographical Information Systems in Libraries" *Database* (December 1994) : 35.
6. Stanley Arnoff, *Geographical Information Systems : A Management Perspective* (Ottawa, Ontario Canada : WDL Publications, 1989) referred in, Allan B. Cox "An Overview to Geographical Information Systems", *The Journal Academic Librarianship,* 21 No. 4 (July 1995) : 237.
7. William P. Kane, "A Hyper Card Call number direction: Using stacks to find stacks", *College and Research Library News,* 50, No. 7 (July-August 1989) : 576-577.
8. Gary W. North, "Will your Library be the spatial data information centre of the future", *Inspel,* 23, No. 2 (1989) : 131.
9. *Ibid.*, 130.
10. *Ibid.*, 132.
11. *Ibid.*
12. ARC/INFO and ARC/View are the GIS software packages brought by Environmental System Research Institute for using mainframe and PC based system respectively.
13. McGramery and Lamont, *op. cit.*
14. Laura Lang, "Mapping the future of map librarianship" *American Librarian,* 28, No. 10 (November 1992) : 880.
15. *Ibid.*

16. Adler, *op. cit.*, p. 235.
17. North, *op. cit.*, p. 131.
18. Cox, *op. cit.*, p. 248.
19. Lisa T. Abbott and Carolyn D. Argentati, "GIS A New Component of Public Services", *The Journal of Academic Librarianship*, 21, No. 4 (July 1995) : 251.
20. *Ibid.*, 253.
21. *Ibid.*
22. Cox, *op. cit.*, pp. 245-248.
23. *Ibid.*, 248.
24. *Ibid.*
25. David J. Cowen, "Importance of GIS for the average persons" in *GIS in Government of the First Federal Perspective Proceedings of theFirst Federal Geographical Technology Conference* (Fort Collions, 10 : GIS World INC, 1994).
26. Abbott and Argentanti, *op.cit.*, p. 251.
27. *Ibid.* 253.
28. *Ibid.*
29. *Ibid.*
30. National Centre for Geographical Information and Analysis NAGIA Core Curriculum, (Santa Barbara, CA : NCGIA UCSB, 1990).
31. Karl Longstreth "GIS Development Staffing and Training" *Journal of Academic Librarianship,* 21, No. 4 (July 1995), p. 271.
32. Joseph A. Boisse and Mary Larsgaard "GIS in Academic Libraries : A Managerial Perspective", *Journal of Academic Librarianship,* 21, No. 4, (July 1995) : 289-90.
33. *Ibid.*
34. *Ibid.*
35. Theresa C. Strasser "Desktop GIS in Libraries, Technology and Cotts : A View from New York", *Journal of Academic Librarianship,* 2 : No. 4 (July 1995) : 280.
36. Brent Allison, "University of Minnesota : Remote Access and the Internet", *Journal of Academic Librarianship,* 21, No. 4 (July 1995) : 283-88.
37. *Ibid.*, 283-284.
38. Ken Laing and David Mander "Where history and technology meet" *Library Association Records* 98, No. 8 (August 1996) : 420-421.
39. *Ibid.*, 421.

4

Mapping of Library Collection

ONLINE PUBLIC ACCESS CATALOGUE (OPAC)

The process of conversion of traditional card catalogue into a computerised one has given birth to Online Public Access Catalogue (OPAC). While referring to the card catalogue, the reader approaches one particular drawer which is relevant to his search out of hundreds of drawers in the card catalogue cabinet, involving considerable physical exertion in order to look for the particular entry he has in mind. While using the OPAC, the reader interacts with or enters the search query through a keyboard of the computer, which executes the search and displays the results on the computer screen. In using the OPAC the constraints faced by the reader in using card catalogues are automatically removed. Moreover, the computer interacts with the reader through a set of questions further to accurately aid the reader's search in the light of the query entered by the user. Computers with OPAC not only interact with one user at a time but also with many readers simultaneously by using its distributed processing capability and give each user facilities as if the entire catalogue is available at his/her disposal.

First and Second Generation OPAC

Subsequent to initiation of library automation process, which began in 1954, the Online Public Access Catalogue (OPAC) came into existence in 1974[1] and became universal in the beginning of 1980s.[2] Thereafter OPAC has evolved remarkably and due to its constant development newer versions of OPAC are made available. These developments have been tracked down through three generations, as categorised by Hildreth.[3]

According to him the first generation of OPAC was just a mechanised version of the card catalogue, whereas features of second generation OPAC are listed as below in Fig 4.1, as described by him:

- "Action" and "how to" option prompts
- Boolean searching
- Full standard bibliographic records
- Help facility, context sensitive
- Indexing term browsing
- Informative error messages
- Interactive search refinements modifications
- Keyword access
- Multiple display format
- Search results display/print manipulation
- Search term approximate match routines
- Shelf list review/scan
- Subject access via library-assigned headings
- Two or more dialog modes."

Fig. 4.1 : Features of Second Generation OPAC

Features of Second Generation OPAC Quoted from "Advancing Towards the E[3]OPAC. The imperative and the path.

If the capabilities of the present systems of OPAC are compared with the second generation OPAC as envisaged by Hildreth it is observed that the today's OPACs have not acquired all the above features and one has to wait for some more time to see them in operation.

Third Generation OPAC

Development of OPAC ought not stop even after all the above features are implemented. It is estimated that the need of OPAC will sustain until 2006[4] and it is going to develop and innovate further. Researchers studying OPAC, are putting their efforts to make the OPAC a "Full Collection Access Tool" within the library and try to facilitate extended access to resources of worldwide libraries too. Hildreth has specially named the third generation OPAC as E^3OPAC where E^3 denotes Enhanced, Expanded and Extended.[5] While describing the features of the third generation OPAC he has listed the features as given in Fig. 4.2.

- The second generation functionalities (Listed in Fig. 4.1)

Plus

- Natural language query expressions
- Automatic term conversion/matching aids
- Closest, best match retrieval
- Ranked retrieval output
- Relevance feedback methods
- Intelligent navigation aids
- Integration of keywords, controlled vocabulary and classification based search approaches
- Expanded coverage and scope ("the full collection access tool")
- Extended access range via linkage and networks

Fig. 4.2 : Third generation OPACs : Outline of function enhance-ments, quoted from "Advancing Towards the E[3]OPAC imperative and the path[3].

The above mentioned discussion reveals that when the developments in the fields of telecommunication, computer science and the future needs of the services of libraries are taken into account all the above developments of OPAC appear to be logical and possible. This means that there is tremendous scope for OPAC to develop further. Ideally speaking OPAC which may look today as a dumb terminal would be replaced by intelligent terminal providing the readers all the benefits that are offered by the micro-computer and making the "OPAC" a "Magic window" to the information sources available within and outside the libraries.[6]

Swanson's Visionaries about the Future Catalogue

Don Swanson undertook an important research study in 1960s with the support of National Science Foundation (NSF) at University of Chicago. The aim of the research study was to find out requirements of the future catalogue. The findings of the project were summarised by him in 1972 and announced subsequently.[7] However, Swanson is best remembered for his article published in 1964 entitled, "Dialog with a Catalogue".[8] This article and the project findings have reflected his views about the future library catalogue. This physicist and Dean of Graduate Library School of University of Chicago had predicted the emergence OPAC of twenty-first century some thirty years ago. His study has generated many research opportunities. Some of the features recommended by him have not been implemented. Although a few of them have been rejected as unrealistic, the desire to implement others have been expressed.[9] When the idea of OPAC was only in infancy, Swanson discussed the possibility of providing access to parts of a book. The purpose of the present study here is not to discuss his remarkable vision in detail, but to highlight some of their features with reference to the topic of the research. It would be interesting to list some of these functional performances of OPAC as imagined by Swanson which are still to be implemented.

1. Catalogue access via microform (now it could be via CD-ROM)[10] back up, facilitating browsing book images. Technologically it is feasible by using high resolution work stations (for example Next)[11], which can present the picture of a book in exact colour, size and texture plus a spindle capability to show the images of book from various angles. There exists a CD-ROM based bibliographical source entitled "Book Scope" brought out by a UK based company in 1996[12] which in addition to conventional bibliographical features the CD facilitates to browse through summary, contents, the author's short biography and cover jacket clippings also. it confirms that it is now comparatively easy to have enhanced non-conventional bibliographical details for the use of readers through OPAC. While suggesting these features Swanson had in mind that the OPAC could be used as window to browse through the library collection seated across the computer terminal.
2. Provide search to the readers through four major components, such as :
   (a) Specific work,
   (b) Subject selection,
   (c) Similarity Selection, and
   (d) Previous use, in order to satisfy the variety of bibliographical needs as well as to utilize the subject expertise available with the reader. These recommendations were one step ahead what we see in the latest commercial search software that comes along with CD-ROM databases or available with library catalogue to search data.
3. In adding information other than the conventional or traditional bibliographic details such as title page, table of contents and index of any work, the purpose Swanson had in

mind was to provide the reader with comprehensive and manageable research results so that his search would be positive, specific and result oriented.

4. Mass storage of full text data for an entire library's holdings, adds new features of document delivery either by on screen display or by paper bound print form to the OPAC capabilities. With the cost of magnetic storage having reduced down, the idea of having comprehensive collections stored electronically on computer and making it feasible to search each and every document available in library through OPAC and also physically having it is not impossible today.[13]

OPAC : The Future Comprehensions

Out of Swanson's vision referred to above, some have been criticised as old, obsolete or unnecessary. However, many of the others are accepted as feasible but are yet to be implemented. Researchers are working on them and wish to make these possible. As such it confirms that the OPAC has a lot of scope for further innovation. In addition to what Swanson has imagined in the past and what has been recommended by others, what has been discussed in this chapter, there are many more things which could be introduced as additional OPAC functions such as : Making use of multimedia technology in order to make human OPAC interaction more meaningful. As suggested by Lee[14] multimedia can be used in improving research retrieval or in making search more interesting by showing 2D and 3D images of objects such as card catalogue, drawers and bookshelves. It could be so reinforced that reader enjoys it as if he is using the library physically although he is using or handling just the data, through OPAC.

If all the above features including Swanson's recommendations are actually implemented since they seem technically viable, then OPAC could provide an entirely new potential to catalogue access tool and these additions would be an advanced type of retrieval feature. These advances can give a new vision to the readers in fully exploring library resources as a microscope of the library collection.

Even with the achievement of the above capabilities OPAC must continue to develop further as there will be many researchers and scientists like Swanson engaged in further research on OPAC. A feature like the Mapping of Library Collection would be an altogether new idea to incorporate, in the OPAC capability which the researcher wants to discuss in this chapter. The idea would be similar to navigating for the required library material in a simulated view of the library stacks in a digitized floor map.

Mapping of Library Collection

So far, OPAC only facilitates viewing bibliographic or catalogue data with all its advanced capabilities. The mapping of library collection using Geographical Information Systems (GIS) can add the additional capabilities of handling geographical or location data of library material, in addition to existing data to OPAC. In that case one can also see the physical location of the library material on library floor maps. However, there is a need to call for some of the features recommended by Swanson to supplement the idea. The proposed system in the project requires some of the following non-traditional bibliographic details such as :

1. Full bibliographical details along with location data.
2. Entry of each and every document available in the library to the library database.
3. Images of book in different angles.
4. Table of contents.
5. Preface.
6. Jacket cover clipping.
7. Index.
8. Locational data of the library.
9. Map data which refers to the location of reading material available in the library where interaction of the same is perfectly synchronized with physical arrangement by linking bibliographic and location details.
10. Multimedia or animation effects to be used to allow user to travel in the simulated library by just sitting at a computer terminal.
11. To gradually achieve the full text information of all documents which are more in demand in the library.
12. Linking of queries with computerised document delivery system.

LOCATION DATA OF LIBRARY MATERIAL

The library contains various forms of library material and each of which carries significant location details. So far these locations are hidden in call numbers and the same are imagined in terms of actual location details such as stacks, row, rack, shelf numbers etc. by the reader. Each document sitting at a particular place on the library shelf, has the definite location details, as remembered by the reader. However, they are not recorded in the present library system and there is no mechanism provided therein by which readers can find out the exact location details.

How Specific is the Location of Books in the Library ?

When OPAC was first introduced it was just a replica of the card catalogue. It did not serve any more purpose other than facilitating search through a computer. Subsequent research has aimed at innovating further and has resulted in the emergence of successive generations of OPAC as discussed earlier. The trend now is to make OPAC a comprehensive, convenient and more user friendly search facility. Further research focuses on optimally incorporating information like location data and circulation status. However, location information which is provided is very broad such as particular library, particular floor, particular area or stack as shown in the screen display of 'Infotrax' system at Rensseler Polytechnic Institute Troy, New York[15] where such information is collected in one of the MARC fields, and the same is converted into location statement. As far as incorporating the precise location of library material is concerned not much efforts (research) have been made. In the process of finding location of books in the library, readers have to make mental efforts while interacting with catalogue and the actual shelf arrangement, which results in finding out the exact location of the material. In doing so the reader has to understand how the library has classed the items, how it has arranged them on the library shelves and finally he has to interpret the special codes in terms of actual location.

When so much is done to improve OPAC then why can OPAC not give the location plainly?[16] As far as specificity of location of library material is concerned the behaviour of the readers itself indicate that the locations of the library material in the library is important. The readers, once successful in finding the location of material, prefer to go to the shelves first, in the subsequent visits. As indicated by Line giving reference to Parry Committee findings that 92 per cent of undergraduates and 72 per cent of staff and graduate students consult shelves first.[17] The readers prefer to go to shelves first also because they are sure and confident about location of the material on the shelves. The characteristic of specificity of the shelf arrangement gives them the confidence. There is no reason why there should be change in the location of the material; else the task of finding the books will be still more difficult.

As long as books are arranged in a classified way, classification scheme controls the arrangement (i.e. in creating order of class numbers as numerals and other order recommended in the scheme). Whereas within one class number, where many books having same subject are brought together, books are further arranged alphabetically by the authors' surnames. At any moment taking into account the above arrangement the position of each book on the shelf becomes unique and it remains static for a long period of time.

Since the location of the library material is specific it is reasonably possible to show the exact location in the OPAC displays. The possibility of showing exact location is stated by Stephens[18] in these words "in fact, because library shelving is quite immobile, it is not unreasonable to expect OPAC displays to show the precise locations of each item in the collection down to the specific range stack and shelf number". The characteristics of specificity of library material reasonably also holds true even in the situation of the growth of the library collection and weeding out of library material.

Undergrowth of Library Collection

The fifth law of library science enunciated by Dr. S.R. Ranganathan is "Library is growing organism".[19] The library is growing but it is also a fact that it is not growing as it was growing in the earlier days. Almost static library budgets, coupled with general inflation and price rise specific to books and journals have considerably brought down the growth rate of the library. There are several other factors like weeding and discarding; microfilming and digitizing full texts and implementing recommendations of Alkinson[20], to limit the growth of library, which help further in controlling growth of library collection.

Only one factor, which may result in changing position of the book from one shelf to the another, is the growth of collection due to limited addition of books to the library. However, there are several other factors which try to partially nullify the growth of books caused due to addition. These factors are listed below in Table 4.1. along with their rate of limiting collection against the rate of increase due to addition.

Table 4.1 : Growth and Reduction Factors affecting Library Collection

Reason	*Addition*	*Reduction*
1. Addition of library	1.89%	
2. Difference of Books issued out of library and returned back (4 .4 - 3.17)		1.23%
3. Books lost by the readers (only partially replaced)		.004%
4. Books stolen by the readers (only books which are in more demand replaced)		.0007%
5. Books weeded out and kept elsewhere*		
6. Books mutilated		0.1%
Total	1.89%	1.3347%

* Action of weeding of books is yet to be initiated.

The above statistics are taken from the circulation activity of a large university level library working in the area of Science and Technology having a collection of 1.80 lacks books established 39 years back. Anonymity is maintained here in agreement with the concerned authorities. The Table 4.1 shows that the growth is about .5653 per cent which affects the arrangement on shelves of a large library very marginally.

It is also true that the addition of books is not uniform in all subjects. For example addition to subjects like science and technology is more as compared to humanities and social science subjects in a Science and Technology library. The standard allowance of 25-30 per cent made in the shelf-space for new additions is thus not maintained uniformly. The variation in the addition of books could be adjusted by organising the expansion allowance such that frequent rearrangement of books on shelves would not be necessary.

If one compares the results of additions in and reduction from the library collection due to factors discussed above, it is revealed that the library collection is not growing as it is generally claimed. The fact, therefore, supports the above statement that the position of a book on a particular shelf is quite immobile or remains constant for a considerably long time.

Location of books when books are weeded out and kept elsewhere

When the books are removed from their normal position and are kept separately the location aspect becomes more important. When the books are removed and kept in compact storage they need to be arranged in the same way as per classification number, or alphabetically by title or arranged serially. The position of the book becomes more unique since the rate of occasional addition and removal controls the growth in that collection. The arrangement of such a collection remains more or less static unless major withdrawal of library material or weeding out is taken up.

Role of Classification

In the process of classification and arranging books on the shelves in a classified manner books on like subjects automatically come together. This helps the readers to browse through the collection. The class number as a part of call number also serves as the location number in the library which is consulted by the users in locating books on the shelves. The call number serves as the linking device between catalogue and the shelf arrangement as the readers retrieve the same by referring to the catalogue before proceeding to locate the material.

The classification scheme once chosen for use in the library remains the same forever unless the library takes a major policy decision to change the scheme. In the course of time developments take place both in subject organisation as well as in classification scheme, which are inter-connected. However, the practice of classification remains more or less unchanged, in order to maintain the uniformity. This refers to use of a particular edition of classification scheme.

Most classification schemes are continuously developed and revised in order to reflect the developments of various fields as well as to accommodate new and emerging fields and subjects. The schedules of revised class numbers are brought out and made available to the classifier regularly. All the revisions are meant for guidance as well as for use and implementation in day-to-day classification so that the resulting advantages are made available to the users. However, it does not happen in the real life situation. The actual scheme used for day-to-day classification is always an age-old edition. This becomes necessary because there are several difficulties in incorporating changes in classification as per revision brought out by the scheme. It involves a huge reclassification and recataloguing work and, therefore time and money. Similarly, there are also several difficulties in using class number (as a part of call number) as location numbers especially where expanded version of class number is used and when book bears a long class number due to complex or interdisciplinary nature of the subject it deals with.

The role of classification could be defined in two ways. From the point of view, designers of classification schemes who take a lot of efforts to comprehend optimum knowledge organisation, it is the subject grouping tool. Whereas in Classifiers' views, who has to introduce a lot of compromises in implementing the scheme, it is a partial subject grouping tool and also a semi-location indexing tool.

The terms semi-location indexing tool is used because of the fact that the call number does not give the actual location number to the reader, but the reader has to interpret and convert it into the actual location number.

Role of Classification as Subject Grouping Tool to facilitate browsing

A Classification scheme should basically function as a subject grouping tool in a real sense, and should try to bring material pertaining to one discipline together for the readers to examine. This facilitates readers to browse through the collection. This function of the classification scheme is well received. However when the same scheme is used for location indexing also then the real effect of subject indexing diminishes as a lot of compromises have to be made in using it as location number and the latter requires broad classification.

Assigning detailed classification numbers especially in number building type of schemes (like Colon classification) brings a lot of practical difficulties in using it as location number. In the collections which are classified by traditionally used classification schemes, browsing is possible with only broad classification number. However, browsing could be more meaningful if classification is done in detail (not broad) by incorporating all the expansions provided by the scheme. Broad classification means a large number of books accumulate on a single class number. The reader, therefore, finds difficulties in browsing when he tries to look for minute subject aspect while browsing. No study seems to have been undertaken to observe advantages and reader behaviour in the environment of detailed classification, as broad classification has generally been the accepted situation because of non-availability of a solution to linkage problem. The present study attempts to provide the solution. In fact there is need to have research favouring this situation. The above discussions reveal that detailed and minute classification would enhance browsing facility further making it more specific and precise.

Role of Classification as Location Index

The library faces many difficulties in maintaining the classification up-to-date following revisions of the scheme. It is natural that reclassification as per latest revision of the scheme is likely to result in long class numbers especially in the case of analytico-synthetic scheme which tries to classify books of interdisciplinary nature. There are practical difficulties in incorporating long class numbers on the spine label of the book. In view of the above difficulties reclassification work is always avoided. Secondly, there is a strong force in the librarian community which advocates the use of classification more as location indexing tool than as a subject grouping tool. The classification is, therefore, maintained broadly to suit the location number. Maintaining broad classification by age-old edition of classification scheme means to sacrifice many advantages made available in the course of development and revision of the scheme. The classification scheme has limitations and remains more of a location indexing scheme and not truly a subject indexing scheme.

Call Number and Location of Library Material

Information about the physical location of library material is of utmost importance both to readers as well as to library staff. In traditional library systems call number including class number, author mark and accession number helps readers in locating library material. The reader therefore is expected to retrieve the same from catalogue in order to find a required book in the library. As such, the system expects the reader to remember the classified arrangement and compels him to interpret the probable location of the particular call number. In case of a large library especially one in a multi-storeyed building the readers need to depend on the library staff for knowing the actual location of the library material as there is neither the mechanism to find out the location of material nor is it mentioned in the catalogue. The library tries to fill this gap by providing guides and signboards. However they find such devices inadequate in finding actual locations.

Location of Library Material as a Part of Call Number

As seen earlier classification scheme performs two major functions such as classified arrangement to facilitate browsing and at the same time combining class number with some other details it performs as a location number also. Moreover, for most of the the time, classification number is truncated to make it suitable for location number. As such it limits the provision available in the classification scheme. It is, therefore, argued that it is not essential for the library classification scheme to perform both the functions of location number and subject grouping as these two factors could be dealt with separately. It is suggested that the system should allow the classifier to classify the material freely. The library can decide upon the policy whether to follow a system of detailed classification or broad one, depending upon the type of library, level of readership and the size of collection. If the library readership comprises of serious researchers detailed classification could be convenient as grouping of minute subject could also be possible.

In fact no classification scheme has the capacity to entirely cover the location aspect of the library as the organisation of library material differ from library to library. It is the task of the library management to handle this issue. The material is organised in each library in such a way that the available space is fully and properly utilised. Moreover the location aspect needs to be handled in two different ways: first, the locations of different types of material within the campus or within the building; and secondly, the location of a single book in the library upto a rack or shelf level. In the present situation where the computer is available at the disposal of the library staff, it is not difficult to spell out the location of the material in the library upto floor, row, rack or shelf number. Moreover, it could add one more user-friendly feature to the OPAC.

Location Data tracked separately as Data Element

Before the computerised library system came into existence, there was no mechanism to find out the exact location of library material, in traditional cataloguing. However, using a computer, it is possible to provide the reader with the exact location of a book at the end of browsing of OPAC. The system, therefore, requires an input of the location data in terms of floor, row, rack and shelf number of each item through added fields. Once the system knows the location of library material things become simpler for the user as it would be possible to provide exact location of the library material in OPAC retrieval. The system, therefore, allows classification to be used to facilitate browsing to the desired extent of classification on one hand and it also informs the users about the exact location of material, without affecting the quality of classification.

Significance of Location for Users

All the systems and services are implemented in the library keeping readers, needs in mind. The readers are mainly concerned with finding out reading material without involving in complex procedures and without spending much time. The location of library material, therefore, is significant to the users.

Need

In a large library knowing the location of library material organised on various floors is time consuming. It is impossible to provide assistance to each and every reader. He is left alone and expected to have mastery over the library system. As far as utilization of library collection is concerned this is not a happy situation. So far there was no mechanism available in finding out the locations to help the readers. Their interaction with the library will be easier in the library if OPAC comes forward to help them. They will definitely appreciate any measures taken by the library towards information of the exact location of library material.

Kane has expressed his concerns that quite a large number of users generally face the general question "How do I find a book in this library by call number"? He observed that readers always go for the short cut method by asking the staff the exact location of material or would jump to the last step of the system which provides the location (Kane invented the Hyper Card Call number directory systems to find out the location of the material in the library.[21]

Cataloguing Measures

All the recommendations made by Swanson are centered around the users and their needs. OPAC, therefore, cannot be just a mechanised version of card catalogue. All the useful and feasible recommendations of Swanson need to be implemented particularly into OPAC. Along with all the bibliographical and other details measures also need to be taken to incorporate location details in the OPAC.

Library Shelving Model

As stated earlier, it is possible to record the locational details (text form) in the data base and retrieve the same in a OPAC alongwith a call number and all other bibliographic details. This will be definitely a provision of additional features for OPAC. It is also stated that knowing class number/call number, it is possible to search the location details using intelligent systems.

If we look at the entire process, from addition of books to shelving them in racks in a computerised library, it is a combination of manual and automated systems. The classification, decision about broad location and sequence depending on type of material, maintaining card catalogue and shelving of books are all manual processes. Data entry, cataloguing and retrieval through OPAC are automated processes.

If the above referred system is considered then there needs to be maximum synchronisation of shelf arrangement reflected in the system and the arrangement actually on the shelves. It would be ideal situation where everything is automated to the extent that robots file the books and remove them from the shelves. However, in practice it is possible to automate and generate shelving sequence by using the computer. Primarily, finding out the location details after actually seeing the books on shelf and providing to the system is laborious task. Secondly, new books are added to the collection and books are also removed from the shelves and finding out these changes and correcting the data would be further complicated. The system, therefore, would, not have control over the location data. It would, therefore, be advantageous to have a system which provides a mechanism to find out the location of library material on its own taking into account the above variables which can be provided in OPAC as well as in providing guidance for shelving. It is, therefore, suggested to have an automated shelving model which would supplement OPAC.

The proposed automated shelving model monitors the shelving activity except actually keeping and removing books from shelves. As per the preparatory step, the proposed system needs to know the shelf space available in terms of units of shelves having standard sizes, thickness of books kept there, broad locations (floor and stacks), which sequences are kept where and the type of material. The system also needs to know the classification system alongwith signs, symbols and various auxillary units so also sequence of classified arrangement.

The system starts acquiring data immediately after classifying the books. It acquires data through interactive query system and gets feedback such as classification number, author mark, broad locations, sequence and type of material. The system here needs to have a little higher level of intelligence and logic supported with necessary algorithms.

The system in addition to the above referred details is also told about conditions of flexible and optimum shelving limits, on the shelves.

Since the system knows the shelf length, thickness of books, flexible optimum limit of shelving, classified sequence and other broad locational details at any particular time it knows details of the groups of books kept on any particular shelf with exact sequence by which they are arranged there. The system would, therefore, have data comprising of *'n'* number of sequences, corresponding to *'n'* number of shelves of different sequences and held at different locations. The system takes into account the addition of books and removal of books through its interactions with acquisition and circulation systems; and rearranges data accordingly and tries to maintain the system.

For a given class number, author mark and accession number i.e., call number, the system tries to match with different sequences and helps in giving information about the exact placement of books in terms of floor, stack, row, rack and shelf to the users concerned in text form.

Thus the system would be in a position to provide the location details in all situations like while adding the book to library, while locating books, after removing the books for issuing out and even in the absence of books.

It may be, however, noted here that so far the actual geographic features in terms of library floor maps are not considered. These aspects are discussed in the next section on the use of Geographical Information System in Mapping Library Collection.

It is concluded that with the automated shelving model as discussed above the system can be in a position to monitor the shelving activity pertaining to the entire library collection. It can give appropriate guidance to the library staff responsible for shelving in knowing the exact place of the newly added book to the library, so also exact arrangement sequence of books may be maintained on the shelf. The required rate of shelving allowance, and fixing the flexible and optimum limit would be guided by the rate at which books are added to that subject. However, this requires systematic study of books added to the library in various disciplines.

USE OF GIS IN MAPPING OF LIBRARY COLLECTION

It is very difficult to predict at this moment the form and type of material a library will have in future. When libraries entered into information age, and when information explosion made its impact, it was predicted that future libraries would be "Paperless Libraries" and future information products would be only in the micro form or electronic form. Although libraries today have access to online information resources, they do contain off-line products such as CD-ROMS in their collections; moreover they also have some full text publications in electronic form. However this does not yet confirm that these will be the only products that would be published and added to the library. The print-on media has still remained as the major carrier of information. It seems that print-on media or books, journals and reports will continue to exist for a considerably long time. Librarians, therefore, will have to continue with their endeavour of processing and disseminating information available in printed books, journals, etc., through their traditional tools.

However, a definite change is seen in the physical form of these tools, for example, card catalogue is changed to electronic form. The day is not far when the classification schemes would be made available online. DDC 21st edition is already available on CD-ROM in a network version, which has the facility of number synthesizing as well as intellectual and material mapping between LCSH and DDC subject headings.

The library catalogue which has been converted into electronic form called OPAC is not just mechanised version of card catalogue but is also further developed and innovative. Now it gives maximum book details. It can also give broad location and circulation status as provided in 059 MARC field which is retrieved from the MARC 050, 090 and 099 fields.[22] It is clear that OPAC's potential is unlimited. Several new features could be added to it including one called "Mapping of Library Collection" as considered in the present study.

The concept of mapping has been associated with the library in various contexts. The concept is implicit in the existing services. So far mapping has been discussed in the context of describing literature of a particular author or literature on a particular subject, which is stacked in the library. This type of mapping consists of a well arranged list only in text form. Pathfinders in various subjects could be categorised in this type of mapping. The concept moved towards being more explicit when it was discussed in terms of text as well as graphics to describe the information resources in a compact atlas form termed "information highways". The idea of this type of mapping has been discussed by Murr[23] and others particularly to describe, "information delivery networks". The purpose of the authors was to summarize the use of information/ telecommunication technology, to understand them and to describe the public information system in use in the area. They have designed a compact atlas as model. The atlas combines data, information blocks and maps which show the actual location points of information delivery systems. The model utilizes maps, texts and graphics to create a pictorial view to conceptualize what exists and what might exist.

The mapping model where maps, graphs and locations are combined with textual information becomes more informative and helps in picturising the concept/view of any presentation or idea better than a model where only texts such as lists are used. In order to describe different analytical features of such a model, one needs to prepare an atlas of such hypermaps. Secondly, such models could be more meaningful if they are made interactive and online through the use of computers.

At a practical level, this type of model is seen being utilised in a recent article in Library Association Record entitled "Where history and technology meet". The paper discusses a very interesting window-based software developed to study historical and archival premises. Details of the same have been discussed in the previous chapter. The system has provided the location points its textual details plus the photographic images of that particular location. Use of geographic map, non-geographic details and images in the form of multimedia is very interesting combination.[24]

The mapping concept described in the above models is a generic one. The model which the researcher wants to present is a specific type and it is expected that it will help in visualizing a similar situation when applied to the library.

By using mapping concept it is possible to prepare a model for describing locations of various items of accessories and spare parts located in a multi-storeyed building of the warehouse of a large automobile company.

Maps of different floors could give an idea about the types of items/spare parts kept at particular places on respective floors. A map could be supplemented with text information giving description of items such as code number, specifications, size, weight, type of vehicle plus the stock position. The atlas of a combination of such maps also combining textual data could give the complete picture of the entire warehouse in a form in terms of simulated view.

It may not be convenient to describe the actual stock position and changes in the locations as they vary with time. Describing these changes manually in graph and text form is cumbersome. However, the situation could be handled successfully by a computer. A special kind of information system known as GIS which can handle map data and supporting text data simultaneously is needed. With the help of such a system it could be possible to show simulated views of various floors on computer screen with details of what types of items are stored. Where they are stored? and what is the stock position at any particular time? Such a system would be the most efficient system of inventory management for the warehouse of a large automobile company.

The concept of Mapping of Library Collection, discussed in this project, would be along the same lines of discussion as "Hackney on disc" and the inventory management system of industrial warehouse handled by GIS described above.

In the process of leading the discussion to the important and contributory topic, viz., "Mapping of Library Collection" a question comes in mind whether the discussion intends to go backward i.e., from present developed stage of library system to the stage of "storehouse"? The answer to this question is "no" as the discussion has taken into account of all the developments that have taken place so far and touched upon future developments as envisaged by the researchers. It is emphasized that the developments should continue further. The purpose behind introducing the new concept as a part in the development of advancement of OPAC, is to provide a refined search tool or comprehensive key to the library collection. By utilizing the same it is expected to provide ease, utilize library collection to the maximum extent and fulfil all the conveniences as envisaged by Ranganathan in his five laws of library science; and to have an additional aid in searching library material with the help of new technology. In the whole process the researcher has used the static library environment or the store house situation as the starting point or the available platform as one of the developmental stages in presenting a further stage upto the advanced stage as discussed today.

New Approach to Handle Locations Data in the Library

Mapping of library collection using GIS would be an altogether new approach in handling the location aspect of library material in the library. It could be one more feature added to the existing capabilities of OPAC. The system would play the supplementary role to the existing system of classification by providing additional location details without undermining the importance of the classification scheme which provides indirect location details with reference to the classified arrangement.

The concept of "Mapping of Library Collection" in the present study

First of all the floor maps of different floors of the library, including the layout of furnitures and fixtures placed on respective floors drawn are to be prepared. The map would look like a two dimensional aerial view of each floor. These maps are then to be digitized using digitizing technique and appropriate technology viz., Geographical Information Systems (GIS). On the other hand attribute data of the entire library collection (bibliographic as well as enhanced with non-conventional bibliographic data as discussed in para "Mapping of Library Collection") is prepared with additional location code in computer readable form for each item in the collection. Both these data are then linked with location code as common element by using linking technique as provided by the particular Geographical Information System package used for the purpose. With the above preparations the system would work on the following manner :

1. For the multi-storeyed library building physical arrangement of books and other material on all the floors could be visible in simulated compact form on a single screen of the computer with zooming facility to see each part alongwith a brief description to know what is kept and where.
2. A small part within a global view of the above could also be visible separately depending on which form or type of collection like general books, reference books, journals, (current and back volumes), standards, pamphlets, non-book material, CDs, etc is to be located there.
3. The system could have efficient query facility to attend to the following requirements :
   (i) Browsing the library material, in terms of bibliographical data, broadly i.e. from stack to group of books on a subject. The browsing would be sequential from each rack to shelf;
   (ii) Random browsing with a particular subject in mind;
   (iii) Locating material with a particular form in mind;
   (iv) Specific query relating to locating particular book to be answered;
   (v) Locating material in different forms across one floor;
   (vi) Locating book in the floor map with any one of the bibliographic elements in mind;
   (vii) Location to be seen, in first phase, upto to rack level,
   (viii) Location to be seen in second phase, i.e. upto shelf level.
4. If possible multi-media and animation techniques to be added to the system to have real simulated view of stacks, rows, racks, with given pathways and directions to reach the material.

The possibilities claimed above are based on the observations made by the researcher and the knowledge gained by him while designing an experimental prototype of GIS-based library system to present an idea of mapping of library collection. The present discussion intends to provide the theoretical background to the experimental part so that the practical and theoretical explanations would complement each other in conceptualising the idea fully.

The functioning of such a system would require following things :

1. Assumptions,

2. System requirements,
3. Type of bibliographical data,
4. Digitized images of floor maps, and
5. Linking of map and attribute data.

Assumptions

The proposed system works on the following assumptions :

(a) As pointed out in the discussion on the locational concept in the library, the place of the book in the library is significant and also specific. In a large library, although the library collections in general are supposed to be dynamic, due to various factors discussed in para "Location Data of Library Material", it can be assumed that the library shelving or place of book in the library is almost immobile for a considerably long time and frequent major movement of material taking place in a manner of a chain process like nuclear fusion is unlikely.

(b) It is also assumed that the library will use the shelving model systems as described in para "Library Shelving Model" where the entire shelving activity is monitored by the system.

System Requirements

The system of mapping of Library Collection has requirements such as equivalent locational data, enhanced bibliographic data, digital images of floor maps, linking map data and attribute data and suitable GIS technology and software. They are described in detail as below :

Equivalent Locational Data

The GIS based library system, which may handle the actual geographical location details of material requires equivalent location data. It will be the code describing floor number, row number, rack number and shelf number. This actual locational data is fed to the system except "n-th" position of the book on the shelf as it varies from time to time. With little orientation which can be provided on the OPAC/system, it would not be difficult for users to interpret and remember the location details.

The location numbers discussed above would be in addition to classification number/call number as system would expect that books would be arranged thereafter in the classified manner.

Type of Bibliographic Data

The system would expect the normal bibliographic data of each and every document available in the library. However normally in the library, a catalogue record, contains data pertaining to books and monographs only and it does not represent data pertaining to journal articles, pamphlets, reports and standards, non-book material or even parts of corporate book. Since the system would be in a position to handle locational data of each and every document of the library, it is suggested to have data of the entire collection so that the computerised catalogue would have comprehensive and enriched data rather than represent only a marginal collection.[25]

Secondly, it is also suggested that data should have non-conventional bibliographic details also in addition to conventional bibliographical details, such as contents page, preface, index and cover clippings as recommended by Swanson.[26] The non-conventional data is required in the light of the possibility of browsing facility, in terms of bibliographic data to be provided in OPAC as claimed in the study. It is felt that the library need not take any additional efforts to enter non-conventional data as above, as they could be downloaded from the commercial bibliographical databases available in the market by matching the details.[27]

From the reader's point of view, it will not only be beneficial to have an exhaustive database of library material irrespective of type and form and having enhanced details but it would be innovative too. It will be interesting to see the data alongwith location using GIS.

Digitized Images of Floor Maps

The system of mapping of the library collection using GIS requires geographic features viz., map and location data. Map data comprises of floor maps drawn according to exact scale and dimensions with layout of furnitures and fixtures drawn on them. Geographical Information System requires these maps to be digitized and stored in the correct form. Digitizing requires GIS software, digitizer and some technical skills to create a digitized map for the system.

Linking Data (Geographical and Non-Geographical) through GIS

GIS has the capability to not only handle data with efficient Data Base Management System (DBMS) but also to handle the correspon-ding geographical data simultaneously. Thus while answering any query, the system gives output in terms of map (point, line and area) as well as associated non-geographical data. This, of course, requires some relation between these two types of data. GIS does it by establishing a link between them through some element common to both geographic and non-geographical data. In most cases location code or polygon number facilitates linking. For the library database the actual location code number given in the bibliographic data base is also given in map data to identify the area where that book is located.

Limitations

The study at its present level reveals that the system works perfectly well having fulfilled the assumptions and requirements mentioned above. However, the following limitations will have to be accepted, although they can be overcome by further research :

1. Although the study claims to show the location of book upto a shelf level, which is no doubt a great achievement, it cannot claim to show the "n-th" position of the book on the shelf as it requires a higher level of programing in visualizing the available books on the shelf and finding the exact location of a book by the system upto "n-th" place from left to right. However it is stated that logically it is possible to show the same and the suggested shelving model is the solution.
2. The study does not discuss one more aspect of such system viz., showing the simulated view of the exact and live picture of the books available on the shelf at any moment as it once again requires a high level of knowledge of video, computer, online, and image processing technologies.
3. The system assumes that it will have the entire collection on shelves. The items which are temporarily removed will also be present in the system in terms of data although they are temporarily absent from the shelf, (due to say, borrowing or temporary removal for photocopying, etc). While stating this it is assumed that the arrangement status of the shelf at any

particular time would match exactly as system shows, except items physically not present. There could be a small fraction of mismatch because in the system arrangement is done logically by computer, whereas on actual shelves the arrangement is done by humans. Human element is likely to bring some errors in matching the result shown by the system and the actual physical arrangement of books. Wherever humans are involved some allowance is always made a fair compromise in any of such operation. This will also be true in the system of mapping of library collection using GIS. Still, this factor does bring some limitations. The system would be ideal where removal and replacing of books is done by a robot at the instruction of computers. Such a system at present however only seems possible theoretically and not practically.

Benefits

Retrieval Enhancements

Retrieval of required information in the system where collection is mapped by using GIS will be through OPAC as usual. The reader interacts with the system through a set of queries and the system responds to them on screen. The retrieval here, however, would be of an enhanced type. The additional features would be as follows :

1. Seeing location(s) of the material in library in the floor map on computer screen.
2. Seeing the location of books in terms of layers from floor to row to rack to shelf.
3. The system provides browsing facility in terms of data. The system provides browsing at different levels as follows :
   (i) Browsing broadly or overall browsing in different types and forms of library material—more of orientation type to understand what is kept where.
   (ii) Browsing by querying in broad subject field.
   (iii) Browsing by querying specific to a topic in a discipline field.
   (iv) Browsing the works of one author spread in different forms and shelved at different places in the library.

(v) (a) Browsing by querying specific title but with brief

bibliographic details.

(b) Browsing with additional details, such as,

(i) summary of books or other item,

(ii) contents, preface, and index of a book, and

(iii) image of book.

4. Circulation status.
5. Exact Location details in text form.

Use of Multi-Media, Animation and Other Technologies and Future Comprehensions

The study has noted that Swanson idea to use microfilm to store the library's catalogue data alongwith physical features of books. It has also noted the idea of other researchers to make use of CD-ROM for the same purpose. Subsequent developments have indicated to make use of computer graphics in order to bring graphical browsable features in the OPAC. It clearly shows that the trend will lead to use multimedia for this purpose.

Browsing GIS based library system as discussed in this chapter will be further meaningful if multimedia, by making use of Virtual Reality Modelling Language or JAVA programing or by using animation technology. By using these technologies in conjunction with GIS it would be possible to create a realistic simulated view of the library on computer screen.

VIRTUAL BROWSING OF BOOKS

Most of the readers express unhappiness about the difficulties they face in finding the location of required library material as the present classification and cataloguing systems have many limitations and readers spend time in overcoming these barriers. However, readers must appreciate the browsing facility provided by classified arrangement. Physical browsing in open access library systems has many advantages. It helps researchers in scanning the materials and get a specific reference which they earlier could not get and such items could prove to be very crucial in the research undertaken.

Browsing the collection in the library in terms of the physical locations, bibliographical details and unconventional bibliographical details is a slightly different concept discussed in the present study. The term used to describe this type of browsing is "Virtual Browsing".

Virtual Browsing means browsing a collection in terms of information through OPAC and that is done before going to the shelves. Such browsing helps readers in deciding whether a particular document is useful or not especially in the absence of it in the library. A comparative analysis of Virtual Browsing with reference to physical browsing follows.

The browsing with the physical entities of 'books' is always meaningful. However it is only meaningful when the book is actually opened and seen. Browsing by just seeing the spine details on the spine is never helpful and it does not give any more information than the catalogue. Moreover, the catalogue gives more information about a book than the spine information alone. However, one cannot rely on the catalogue for browsing as it does not (at least the traditional catalogue in first generation OPAC) give additional information like content or index plus the image. Browsing of books physically is always advantageous. It is just like seeing a list of grocery items and seeing the actually items in a self-service departmental store. Physical browsing adds factors like visual communication, size, face and many crucial factors which help the readers in deciding whether the item is useful or not. With these limitations the catalogue does not provide alternative or second hand browsing. However if the contents of the card catalogue or say the current OPAC are revised then it can provide an alternative to physical browsing. It is possible to add more information to the catalogue such as image of book, content page, preface, index, etc. If all these features are available then the virtual browsing through catalogue would be possible and preferable. A comparison of physical browsing with virtual browsing is presented in Table 4.2.

In fact OPAC takes an edge over physical browsing by adding one important feature viz., browsing details in the absence of a book on the shelf.

VIRTUAL LIBRARY *VS.* GIS BASED LIBRARY SYSTEMS OR SIMULATED LIBRARY SYSTEM

The concept of a paperless library has been discussed at a theoretical level for quite some time. Advancement of computer and telecommunication technology has further supported the idea. However, the claim of total disappearance of the print media is yet not in sight. The visualization of a library, having its information

Table 4.2 : Comparison of Browsing Books Physically and Browsing them through OPAC with GIS

	Features	*OPAC with GIS*	*Physical Books*
1.	Physical appearance size, image of cover title page	Yes	Yes
2.	Content pages	Yes	Yes
3.	Preface	Yes	Yes
4.	Index or Bibliography	Yes	Yes
5.	Author's details	Yes	Yes
6.	Summary on Jacket	Yes	Yes
7.	Browsing in the absence of books on the shelves	Yes	No
8.	Actually seeing the shelf arrangement	Yes	Yes
9.	Intellectual feel of book	No	Yes
10.	To see the details of parts of corporate book	Yes	Yes
11.	Actually reading the text of the book	No	Yes

sources available only is electronic media which could be accessed from a remote place through network facility under virtual library systems still seems a challenging task.

Virtual Library : A Concept—Virtual library thus, is a new concept and the developments in technology indicate that it would be possible. Virtual library is again a gift or impact of technologies which have given a fillip to new relation between human beings and recorded knowledge.[28]

Firstly, the concept is based on real time system, where a number of users are accessing library simultaneously. A reader feels as if the entire library resources are at his/her disposal. Secondly, on the same lines, it connects the reader to other libraries which means a reader has access to greater information than actually present in the library.

Requirements for Virtual Library Systems : Virtual libraries can facilitate accessing bibliographic data from one library to a number of other libraries. The next phase could be of partial conversion of a document into full text database and bibliographic database. Third phase could be conversion of entire print media to electronic database. In the process of establishing "Virtual Library System" the participating libraries that would be interacting with each other in order to provide comprehensive access to the readers, need to prepare themselves by converting their material databases according to the above phases. Such systems could be fully operational only when all the interacting libraries achieve the third phase where the reader can access entire resources of a library or many libraries through a terminal in the library or from home. To attain the above requirements and to establish virtual library system in practices seems a challenging proposition in the near future.

One needs to wait for considerable time to enjoy such a system. There is a need to have an interim phase and to enjoy this interim phase readers need not have to wait for long time. The phase could be called "GIS Based Library System".

Interim Phase of the Virtual Library System : OPAC would be used as a window to the proposed virtual library system. One of the facilities used in a virtual library system is a visual feature where a "Virtual Stack" based on an extended classification searching system is created. This "Virtual Stack" could allow the users to browse shelves, pick the items, open them to see title page, refer to table of contents, etc. This way one can switch from one subject to other.[29] However, this is regardless of the location of items in the "real" library situation and the system does not show to the reader images of material with their real locational relationship. In GIS based library system attempts are made to present library services and resources through a visual and spatial interface whereas virtual library systems present the contents of other libraries.

The virtual library system has several advantages which might open a new gateway and add a new feature to the future librarianship though we have to wait and see its full development and application in a real-life situation. The purpose here is not to doubt the virtual library system but to invite attention towards the possibility of giving online/analog simulated view of actual physical library[30] through GIS based library system. The so-called interim phase "GIS based library system" could achieve this purpose through concurrent use of various information technologies such as GIS (Geographical Information System) and its interaction with multimedia and animation technology.

The virtual system can open a new dimension to library instruction technology like Virtual University or like use of virtual reality system in distance education. Incorporating and integrating library instructions directly into the classroom could be possible by using the same technology. It can also facilitate the library instruction aspect in distance education, which is not possible now, adding further convenience and giving access to information sources.

However, if the GIS based simulated library aspect is added till virtual library technology comes into reality it will not only give readers the satisfaction of using vast information resources within and outside the library but it will also facilitate in making the reader a more independent library user.

CONCLUSION

Some thirty years ago Swanson had expressed his ideas of the future library catalogue viz., OPAC. Many of his visions are yet to come into reality. He had thought of user friendly catalogue system through his ideas about the future OPAC and argued that the future OPAC can have unlimited scope of development. Locational information of the library material was another area where not much research has been done. The idea of mapping the library collection which has been discussed in this chapter is an innovative idea to incorporate in the user, friendly feature of the OPAC. The discussion confirms that location of library material is specific. It is very significant in the users view and it is possible to separately track down locational data which can be handled by the computer through a shelving model. It is further suggested that the primary function of the classification schemes is the subject grouping and not the location indexing tool. There is need to separate these two aspects and allow the use of classification in its free form. The study theoretically confirms that the idea of "Mapping of Library Collection" is feasible. It further suggests that the use of GIS technology in library management and the idea of using GIS in developing inventory management tool in the library can throw open these innovative applications of GIS in other areas too. Many authors have expressed the idea of the possibility of incorporating exact location data in OPAC.[31] Kane had invented the Hyper-card call number directory for finding location of call number in the library.[32] Grimshaw[33] has expressed the possibility of using GIS for this purpose. The discussion on mapping of library collection and using GIS for the purpose is one step ahead in this direction. The study has mentioned many advantages of the system. Virtual Browsing of Collection and browsing books in their absence is one of the significant advantages. It has been suggested that the entire collection of the library with its addition unconventional bibliographic details be incorporated in the system. The system with these additional features can promise the readers a gifted tool in using the library's resources. All these features could be made available to the users through OPAC which is the significant part of the findings of this research project. The study has suggested that the GIS based library system or simulated library system be considered as an interim phase of virtual library system as the latter will take quite some time to be fully developed and functional in reality. It is felt that the attempt made in this chapter to discuss the altogether new concept of "Mapping the Library Collection using GIS" has further outlined and justified the purpose of this research study. However, this was mainly a theoretical presentation in the absence of any evidence of availability of such systems. The next step of this research project is to provide substantiation by demonstrating the idea through the development of a prototype which is discussed in the next chapter.

NOTES

1. Charles R. Hildreth, "Online Public Access Catalogue : The User Interface (Ohio: OCLC, 1982) Quoted in Shiao-Feng Su "Dialog with an OPAC . How visionary was Swanson in 1964?, *Library Quarterly,* 64, No. 2 (1994) : 136.
2. Nancy J. Williamson, "Subject access in the online environment", *Advances in Librarianship,* 13 (1984) : 47-97.
3. Charles R. Hildreth, "Advancing towards the E^3OPAC : The imperative and the path", in Think tank on the present and future of the online catalogue : *Proceedings occasional paper, No.9.* (Chicago American Library Association on, Reference and Adult Service Division, 1991) : 5-19.
4. Nancy J. Williamson, "Is there a catalogue in your future? Access to information in the year 2006", *Library Resources and Technical Services,* 26 (April-June 1982) : 22-35.
5. Williamson, *op. cit.*
6. Gary A. Golden, "Taming the unfriendly system : Microcomputer as patron terminal to access an online catalog" in F. W. Lancaster (ed.) *What is User Friendly* (Urbana : University of Illinois, 1987) : 61-79.
7. Don R. Swanson "Requirements study for future catalogues", *Library Quarterly,* 42 (July 1972) : 302-315.
8. Don R. Swanson, Dialog with a catalog", *Library Quarterly,* 34 (January 1964) : 113-125.
9. Shiao-Feng Su, "Dialog with an OPAC : How visionary was Swanson in 1964?", *Library Quarterly,* 64, No. 2 (1994) : 131.
10. Steve Lambert, CD-ROM *: The New Papyrus* (Redmond Wash : Microsoft, 1986) Quoted in Su, *op. cit.*, p. 155.
11. Su, *op. cit.*, p. 152.

12. Bookscope (CD-ROM), (Woodbridge : Primary Source Media, 1996).
13. Swanson, "Dialog with a catalogue", 114.
14. Newton S. Lee, "Multimedia visualiser : An animated objective based OPAC", *Information Technology and Libraries,* 10 (December 1991) 297-310 Quoted in Su, *op.cit.*, p. 145.
15. Irving E. Stephens, "Getting more out of Call Number : Displaying holdings, locations and circulation status", *Cataloguing and Classification Quarterly,* 13, No. 3/4 (1991) : 99.
16. *Ibid.*
17. University Grants Committee, Committee on Libraries, Report HMSO 1967, pp. 213-228. Appendix 5, Report by Birmingham University Library on Surveys carried out in 1964 on the use of the library by undergraduates, graduate students and staff referred in Maurice B. Line and Philip Bryant, "How Golden is your retriever? in Lines of thought : selected papers of Maurice B. Line (ed.), by L.J. Anthony (London : Clive Bingley, 1988), p. 13.
18. Stephens, *op.cit.*
19. S.R. Ranganathan, *Five Laws of Library Science* (Bombay : Asia Publishing House, 1963).
20. University Grants Committee, *Capital Provision for University Libraries* : Report of the working party . Chairman Richard Atkinson (London : HMSO, 1976).
21. William P. Kane, "A Hyper-card call number Directory Using stacks to find stacks", *College and Research Library News,* 50, No. 7 (July/August 1989) : 576 577.
22. Stephens, *op. cit.*
23. L.E. Murr, J. B. Williams and R.S. Miller, "Mapping Information Library Networks : the objective, the methods, the benefits and the model", *Library Hitech* 4, No. 13 (Spring 1982) : 51-63.
24. Ken Laing and David Mander, "Where history and technology meet", *The Library Association Record,* 98, No. 8 (August 1996) : 420-421.
25. David A. Tyckason, "The 98 per cent Solution : The Failure of the catalog and the role of electronic databases". *Technicalities,* 9, No. 2 (February 1989) : 8-12.
26. Swanson, "Dialog with a catalogue, *op. cit.*
27. Bookscope, *op. cit.*
28. D. Kaye Gapen, "The virtual library, knowledge, society and the librarian", in *Virtual Library : Visions and Realities* (ed.) by Laverna M. Saunders (Westport : Meckler; 1993) : 1.
29. Triangle Research Libraries Network Governing Board, "Concept Paper : the future online systems at TRLN", November 27, 1990, Quoted in *Virtual Library,* op. cit., p.29.
30. Jamshid Beheshti, Valerie Large and Mary Bialek, "PACE" : A browsable graphical interface", *Information Technology and Libraries* 15, No. 4 (December 1996) : 231-240.
31. Stephens, *op. cit.*
32. Kane, *op. cit.*
33. David J. Grimshaw, "Geographical Information Systems in Business", Vol. 54, *Encyclopaedia of Library and Information Science,* (ed.) by L.N. Kent (New York : Marcel Dekker, 1994): 187-202.

5

Development of Prototype

BACKGROUND

While working on various aspects of the research topic, viz., "Use of GIS in Library", attempts were made through literature search, to find out whether any research has been done earlier. If not, has any research been undertaken currently. Whether any literature has discussed this new concept anywhere. Or, whether such a system is operational somewhere. However, no literature on Use of GIS in library per say in the above context was traced. It was, therefore, decided to look into the feasibility of development of a prototype in support of the newly discussed idea so that the concept behind presenting the same could be cleared.

The study has discussed various possibilities of using GIS in library theoretically. However, the discussion was concentrated on one of those possibilities being important and contributing to the field viz., Mapping of Library Collection using GIS. A GIS based library system, which could demonstrate the idea of mapping the library collection was in the background of above theoretical discussion.

OBJECTIVES

Ideal GIS Based Library System

Ideally any system relating to the use of GIS in the library as discussed above should work in the following manner :

1. The system should be capable of handling the mapping of the library's entire collection available in varied forms and kept at different locations in a large multistoreyed library.

2. It should facilitate digitizing the floor diagrams of a library.
3. The system should be capable to connecting bibliographical data to library material location data in the respective floor map.
4. If possible the system should be able to store the physical images of the reading material.
5. By using the system reader should be able to get the specific location details of the required library material and it should also be possible to see the location of that material in the library floor diagram on the computer screen in a simulated view.
6. The system should facilitate browsing the library collection through OPAC with graphical and non-graphical data with conventional and non-conventional bibliographical details and particularly in the absence of material in the library.
7. The system needs to create a simulated view of library stacks and other areas by combining GIS based library system with multimedia and animation technology, in order to give the reader experience as if he/she is actually browsing in the library through a computer terminal.

Feasible System

In reality, however, the development of an ideal system as referred to above, though not impossible, is complicated, needs a suitable GIS package (hardware and software), as well as a high level of expertise. Moreover, the application of GIS in the library is an uncommon application and there is no such application software readily available. And to specially design software for this specialized purpose would require a large financial support.

Looking into the above problems of developing ideal system it was decided to develop a prototype which would reasonably give a feel of the ideal system so that at least the minimum requirements are achieved in preparing the prototype.

Purpose of Developing the Prototype

The purpose behind developing the prototype, which would be a trial model or preliminary version of the concepts theoretically discussed in the earlier chapters, is to demonstrate the new idea/concept of mapping the library collection and locating library material in library floor maps through a combination of logical and practical examples.

Coverage

The initial phase of development of the prototype was to basically understand GIS. Subsequent phase, discussed in the next section, would broadly cover the preparations, digitizing of floor maps, creating bibliographical/attributes database, connecting map data to bibliographic data and finally demonstrating the new concepts by various examples.

PROTOTYPE DEVELOPMENT PROCESS: FIRST PHASE

Preparations

Selection of the Library

The GIS based library systems will be useful in any large sized library housed in a multi-storeyed building.

In the light of this requirement the Central Library of Indian Institute of Technology, Powai was found suitable as it has a vast stacking area and is housed in a multi-storeyed building.

Tracing the Floor Map

In order to have the real simulated view of the floor, it was necessary to get the drawings of various floors of the library drawn to scale. For this purpose the concerned departments viz., Estate Office and Design Cell of the institute were contacted. Accordingly floor diagrams of the ground floor, mezzanine floor, first floor and second mezzanine floor were obtained. In the case of the ground floor and the mezzanine floor the drawing of the old building and the annex building were merged in order to have a full view.

Modification and Redrawing the Internal Features

All the above mentioned drawings indicated only the boundaries and it was not possible to know from these drawings where stacking areas, reading areas, office areas, common utility areas, furnitures and fixtures etc. were located. In the above drawings it was really cumbersome to draw all the above features. It was, therefore, decided to draw all the internal features of furniture, fixtures like two dimensional views of library stacks, reading tables, locations of counters, computer room, office, check points, pillars, catalogue cabinets, etc in the ground floor plan only. This facilitated digitizing the diagram so that a clear view of a library floor would be seen on the computer screen as if one is looking at the floor with furnitures and fixtures from the top.

Use of Locally Available Package

There are several commercial GIS packages available in the market as well as those which are developed locally. ARC/INFO is one of the major GIS packages which was available commercially and is the package being used widely around the world. ARC/INFO has been developed by Environmental Systems Research Institute (ESRI), Redlands USA. It was necessary to find out from ESRI if a PC based package could be used in developing the prototype. ESRI promptly responded and suggested the use of their ARC/INFOs front end product ARC/View and requested the researcher to contact their local contact in India in this matter, viz., International Data Management Systems, New Delhi represented now by NIIT. After contacting them they advised the researcher to contact their office at Pune. The researcher then visited them and explained his project and its software requirements. After discussion they were happy to take up the task of developing the appropriate software but requested information of a suitable library and a sponsor who would finance such a project. The finance needed was staggering and impossible for the researcher to raise.

It was, therefore, decided to direct the efforts towards locating a PC-based GIS system which would be easily available and without involving much financial commitment. Some indigenously developed GIS packages like ISROGIS, GRAM, GEOSPACE, GIS NIC, THEMAPS were identified. It was decided to contact one of them. The GRAM software jointly developed by Centre of Studies for Resources Engineering (CSRE) of IIT, Mumbai and DST was easily accessible. Accordingly, request was made to the concerned authorities.

GRAM Package

GRAM is Geo Referenced Area Management Package. It operates on PC and is DOS based. GRAM can handle both vector and raster based data and it also has functionality like raster based analysis, image analysis modules, etc.

Working on GRAM Package

The GRAM package has various modules like Input Module, Attribute Link, Analyse Module, Terrain Module, Image Processing Module and Point Module. It was not necessary to use all of them. The Input and Attribute link modules involve processes like digitizing, cleaning, polygon formation, polygon labelling, and linking of spatial and non-spatial databases. The above processes are discussed step-by-step as follows:

Digitizing

In this process a manually prepared map is transformed into a computer readable form or digitized form through the digitizing procedure. The input module determines *X* and *Y* co-ordinate pairs to describe locations of points, lines or areas as seen on the map. The digitizing tool involves PC having GRAM software which is connected to the digitizing board and the pointer for the input of geographic or map data. The drawing, to be digitized is to be put on the digitizer. All the four corner joints of the frame are to be registered first. The process is called Registration and while doing this a file Test. Reg is created in the system. The scale of the drawing is also to be fed into the computer. With this the digitizing process becomes ready. Using the above tool, all the floor drawings were digitized.

Using this process, drawings of ground floor, mezzanine floor, first floor and second mezzanine floor were digitized. This exercise gave a rich experience in the digitizing process. As stated above the floor diagram of ground floor was concentrated upon for developing the prototype. All the locations of existing furnitures and fixtures were drawn and digitized. As a result of the input of points and segments of the diagram, files such as Test.lin (Segment Features) and Test.pnt (Point Features) were created in the system. The whole process took a considerable long time. It was revealed that the digitizing process is prone to committing many minor mistakes which can create problems in the future.

Cleaning

Whatever errors, that had crept into datafiles during the process of digitization, were corrected following the procedures stipulated in the package. The errors were of the nature of hanging and overshooting of segments. They were required to be removed before proceeding to the next step called polygon formation. The system has a built-in cleaning process by using snap and dangling distance to remove the common errors. But some errors have to be corrected manually. The correcting process is as follows :

A window is to be selected on the map, it is zoomed upon to see the type of error which could be 'hanging, or 'overshooting segments'. These errors are removed by using facilities like move points, insert segment, break segment, delete segment, add point and delete point. After removal of all errors finally the clean programme is run and this breaks intersecting segments at the junction points and updates Test.lin file. In the cleaning process there is another programme called cleanup which removes the duplicate segments.

Polygon Formation

In this process, the entire map is divided into small and big polygons. The polygons are generated with the help of POL-FORM utility where a segment chain is established for each polygon to indicate the polygon boundary. The chain starts with the first segment, moves along its forward link in an anticlockwise direction until it meets the starting node of the first point again. Each segment gets two links viz. positive and negative. The POL-FORM process continues till both the positive and the negative links of all the segments are used by the system for creating the polygon. In polygon formation process corresponding Test. Pol file is created.

Labelling

In this process each polygon is numbered or coded with some unique number as the identification number. The File named TEST File created for this system has over 1600 polygons sprayed over the entire map of the ground floor which was labelled in this process. Each and every polygon has to be numbered in order to proceed to next process viz., Pol-Fill.

Polygon Filling

This process links each polygon with a unique point/name or code and stores data in point file. Point or name or code acts as a linking device between map data and attribute data. After this process the system becomes ready to incorporate attribute database as well, required for demonstrating the results.

Creating Attribute Database

In attribute database, a database is created by keeping the location of library material placed in a particular area in mind. The database can have as many elements. However, it needs to have one data element, which is common to map database and the attribute file also.

Attribute Linking

This utility supports interfacing the digitized map data with corresponding non-spatial data available on a database by using linking utilities. Programs like link and CR-dbf are run and giving the map as well as attribute data files. Finally LN-dbf program is run. LN-dbf links the database with polygon segment and point maps and displays the corresponding attribute information of any polygon.

Achievements in the First Phase

At the end of the preliminary phase it was possible to understand GIS to a sufficient extent, for use in the library environment and to prepare a step by step procedure in the creation of the proposed prototype.

In specific terms a comprehensive library floor diagram was digitized using the available GIS package. The output was used to prepare logical examples of mapping of the library collection. It was a huge task to generate attribute data and demonstrate the concept in the comprehensive digitized diagrams. It was, therefore, decided to stop at this stage. The results are displayed in Fig. 5.1 and Fig. 5.2.

Figure 5.1 shows the plotted image of the digitized library floor map with manually drawn legends showing various locations of the library floor maps.

Figure 5.2 shows how digitized maps of different floors can be

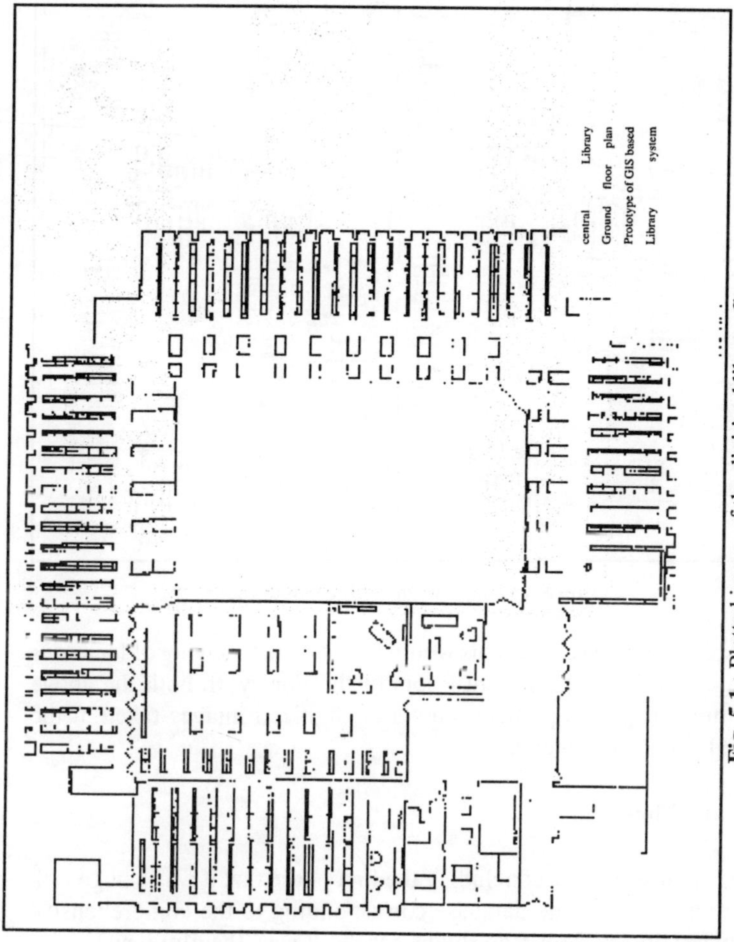

Fig. 5.1 : Plotted image of the digitized library floor map

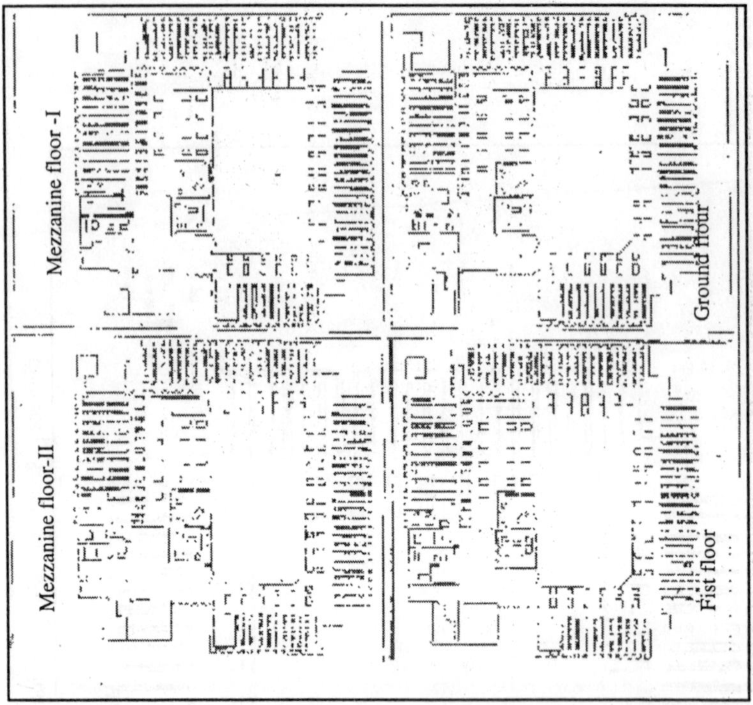

Fig. 5.2 : Digitized map of different floors

combined in one screen. This would facilitate the viewing of locations of material kept at different floors of the library. In both the above examples (figures) plotted outputs of digitized images taken at the initial stage have been used.

Second Phase

It was realised while concluding the first phase that it was a huge task to create the attribute database corresponding to the comprehensive digitized library floor map shown earlier. It was, therefore, necessary to take an abstract version of it and create a small attribute database to demonstrate the concept. In this phase it was expected to see the location of library material using spatial (location) data to obtain non-spatial data (i.e. bibliographical data), whereas it was also possible to see the location of library material using non-spatial data to obtain spatial data.

Here, two layers of map data were involved. One was the floor map in terms of rack dimensions and the second layer was a diagram in terms of shelves of a rack. The system, therefore, enables the browser to go from rack to rack by moving the cursor from one rack to another and allows him to activate the rack further into shelves to see the material kept on each shelf as well as the material kept on different shelves by backward/forward and downward/upward movements of the cursor.

The system further allows the browser to interact with library data through non-spatial data i.e., through any particular field like author, title, subject, keyword, class number, etc. Here the browser is allowed to choose any one of these fields, give the value of that field, get the result in terms of total entries found and allows him to see the details of those entries with right and left arrow keys where as for each entry rack location is highlighted and bibliographic details are displayed by the side. Hereafter, the system was in a position to show the particular shelf where that document is shelved.

For the development of this part of the prototype the same steps are followed as discussed earlier i.e., digitizing, creating attribute data, linking attribute data and map data and retrieving the information.

However, in order to utilize the GRAM (GIS) package for library application considerable programing efforts were required as this type of retrievals was beyond the boundaries of the package. The original package only handles one layer type of map information and shows one to one relation only. After considerable modification i.e., handling multilayer or multirecords type of data was introduced here, it was possible to incorporate multiple information about a number of books kept on shelves and thus two layer type of retrieval was possible.

Digitizing and Other Steps

The floor diagram which was digitized in the previous part was used here but in an abstract form. This functioned as a first layer diagram. The second layer diagram i.e, front side of rack in terms of shelves were digitized. The digitization was completed by following further steps such as cleaning polygon formation, polygon filling, etc. The library floor diagram called Sun and the rack diagram called Moon were used.

Creating Attribute Database

In this case, to facilitate linking of attribute data with two layers of digitized diagrams two databases called master.dbf and book.dbf were prepared. Master database contained elements referring to rack details along with other system data and book data contained all the details of book plus the rack details. The rack and shelf data are used to link map data and attribute data.

Linking Mapdata to Attribute Data

The map database pertaining to the library map diagram viz., sun with its different versions i.e., Sun.Lin, Sun. Pol, and Sun. Pnt, alongwith other system files were considered for linking. The attribute database such as database of rack details called Master.dbf and database containing book details called Book.dbf were considered whereas for the diagram rack/ shelves called Moon was also considered for linking. After running the appropriate programs, map databases were linked to attribute databases. Since this phase also covered retrieving data in two layers as also handling *'n'* number of records and the system was to provide adequate query facility, the task was achieved programatically.

Achievement in Second Phase

With an experimental abstract floor diagram having 20 rack locations incorporated in it and with attribute data having an adequate number of records it was possible to retrieve the following information on the screen by using the prototype :

1. Browsing from spatial to non-spatial details i.e., from rack to non-spatial attribute of racks such as row number, type of material kept in it by movement of cursor.
2. After activating particular rack in terms of shelves it was also possible to browse books available on that shelf sequentially.
3. It was also possible to browse books from one shelf to the other shelf of a rack, i.e. browsing books in a library floor diagram upto a shelf level was possible.
4. In another mode, i.e., from non-spatial (book details) to spatial

(i.e. locations) searching of books available in the library was possible.

These types of search results as displayed in *Appendix II* helped in clearly demonstrating the idea of "Mapping of Library Collection" as envisaged in the present research theoretically as well as practically through the development of the prototype. However, the comprehensive demonstration of mapping of library collection should include picture which will show the locations of entire material on the subject kept at different locations or floors due to their being in different forms or types simultaneously on one screen as shown in the Plate No. 13 displayed in *Appendix II* just before the results shown in the preceding plate Nos. 9-11 i.e., all items shown one after another.

As shown in the Plate 30 of *Appendix II* a logical example has been given which describes how location of a book could be shown upto a particular shelf level.

Though on a small scale, the retrieved information from the system, as discussed in the above stages of the development of prototype gives idea of "Mapping of Library Collection using GIS", and confirms that such application is feasible. The step by step retrieval visuals in the form of photographs of computer screen are provided in the *Appendix*-II. (Plate Nos- 1-30).

The development of a prototype was restricted to the boundaries and limitations of GRAM software due to the constraints of finance and availability. It would have given better results if a better or higher level GIS package would have been used for this purpose.

Application of the System in Real Life Situation

This type of application of GIS to uncommon fields like library management is very new and it will take some more time for the idea to gain full acceptance by librarians so that it could be implemented.

The library where such an application is to be implemented needs to make some basic preparations in order to create an environment conducive to the system. Not much reorganisation of attribute data concerned to library collection is required. However, location data needs to be defined, entered and tracked separately in the respective fields additionally provided. To facilitate this, database of rack details and shelf details are need to be created.

The procedure can be evolved, like stock verification situation where all the books are present on the shelf, the sequence of all books kept on each shelf arranged class number-wise is noted in terms of their accession number sequence. By using this sequence the corresponding data for the respective group of books kept on the particular shelf, is retrieved from the main computerised catalogue database. After downloading the data for this group of books to the modified database which has additional fields to record the location details, the corresponding location details are to be found out and entered. These details basically contain rack number and shelf number along with floor, stack number and row details. The text comprising all these locational details could also be retrieved with bibliographic details. The rack number and shelf number and further details are separately handled in the attribute data which are used in conjunction with the library floor digitized map in order to retrieve the location of material upto the shelf level.

The policy decision may be taken to what extent a collection would be covered for mapping. It is suggested that, as far as possible, the entire collection be considered for the purview of the system. This will enable the user to have control over the information sources available in the library. Naturally it will benefit users in getting information in a consolidated way about material spread in different forms in the library as a result of keeping them at different places.

Attribute data and map data created as above are linked to each other. An appropriate query system is also developed to enable the browser to interact with the menu driven system to see the location of material in the map graphically as well to have bibliographic and other details besides it in textual form. The programing support to facilitate handling multilayer and multi-record environment is certainly required.

The system would be linked to the circulation and acquisition modules through a shelving model as discussed in the preceding chapter. This type of interfacing would facilitate, on one hand, information about the circulation status of a book at a particular moment and in the situation of non-availability of books due to their being issued out (or such other reasons) the user would be in a position to browse through the bibliographical and other details available in the system. On the other hand, it would be possible to locate the placement of books newly added to the library collection. Thereby the system would have full control over the movement of books on the shelves created due to addition and removal.

It is likely that if all the details are invariably shown for each and every search then the system might slow down the retrieval. Instead, the query system could be designed in such a way that all the details are grouped in phases and the user is allowed to choose one phase only as and when it is required to see further details. For material which is not present in the library due to different reasons, corresponding bibliographical and locational details would have to be removed from the respective databases.

With sufficient preparations and provisions made as discussed above, the system would work reasonably well and the expectations envisaged in the research proposal would be achieved.

Cost Factors

The investment cost involves the following factors :

1. Cost of file server of 486 or Pentium with storage space corresponding to the size of collection of the library.
2. Cost of appropriate number of OPAC terminals required.
3. Cost of system development including cost of programing.
4. Cost of GIS package to be used.
5. Actually execution of project, viz., noting down the location details, entering the same in the system, preparing floor drawings and digitizing them plus linking and testing the system.

It is presumed that the bibliographical data has already been created and it can be used with few modifications, if needed.

Conclusion

On the basis of library mapping of library collection implementation discussed above it is concluded that it is feasible to map the entire collection of the library by using GIS and information such as location and bibliographic details. Thus one can very well depend on the system in finding out the location of the library material upto shelf level, and also to browse through the collection through OPAC. It would be a great help and relief to library users. Thus the system can claim a very useful inventory management tool for the library.

REFERENCE

As far as development of Prototype using GRAM package is concerned following GRAM Manual is used : GRAM Users Guide (Geo Referenced Area Management Package Version 2.0). (Bombay, New Delhi : IIT Bombay and NRDMSD : 1993).

6

Summary, Conclusions, Research Findings and Recommendations

SUMMARY

The first chapter broadly discusses all the relevant issues concerned to the research theme, i.e. the concept of GIS, its special characteristics, functions, utilities and requirements. Further discussions include the applications of GIS in traditional and non-traditional areas. Problems in the library environment and the relevance and prospective contributions of GIS in encountering these problems are assessed. The main objective of the study is to examine the way in which GIS can be used in the library. Specific aims and objectives of the study are outlined in the introductory chapter. The methodology applied in the study is also explained.

The review of literature is presented in the second chapter. The sources of literature cover all the materials, first on the broad subject of GIS and then, on its applications to library and information science. The problem faced in the process of identifying the relevant literature due to omnipotency of the subject and acquisition of the same are also stated with purpose.

The third chapter takes a note of the impact of GIS on library and information field. It then outlines various possible uses of GIS in the library and information field. A detailed discussion is centered on the three following uses :

(i) Use of GIS in mapping library collection,

(ii) Use of GIS in providing information services based on spatially referenced databases, and

(iii) Use of GIS in mapping/networking of library resources spread over a city or a region.

The first use as referred above is discussed in more detail in the fourth chapter, whereas the present chapter concentrates more on the second and third uses of GIS. This chapter has also touched upon the applications of GIS in other areas of library management.

The fourth chapter begins with discussions on the emergence of OPAC and its gradual development alongside different generations of computers. It further discusses Swanson's ideas about the future of the library catalogue. Then it concentrates on the discussion of the theoretical concept of mapping of the library collection using GIS. The discussion involves issues such as the significance of location data of library material; the role of library classification in subject grouping, in classified arrangement and in retrieving the material; the need for location finding mechanism; the need for handling location data separately; modelling shelving activity and finally the mapping of library collection using GIS and its benefits. Other related aspects such as virtual browsing and comparison of virtual library and simulated library system or GIS based library system are also discussed in the chapter.

As planned in the study it was felt necessary to demonstrate the concept of mapping of library collection through the development of a prototype in support of theoretical discussion. The fifth chapter therefore, deals with the development of the prototype. The chapter, describes various aspects of prototype developments. They include identification of library, preparing floor maps with necessary location features and identifying suitable GIS package which can serve the purpose. The chapter further deals with the actual development of prototype by following steps such as digitizing the floor maps, creating attribute database, establishing links between map data and bibliographical data and finally demonstrating idea of mapping of library collection and retrieving the location of library material.

The chapter describes the problems faced in developing the prototype viz. in using GIS package for library applications and how they were overcome by modifying the package programatically. The chapter finally presents the results of the prototype in terms of computer screen photographs (appended as Appendix - II) and some logical examples with a view to demonstrate the concept in a practical manner.

An exhaustive bibliography comprising of information sources directly and indirectly related to the topic is appended. Appendices-I and II comprising of information on GIS technology products and the source codes used in development of prototype respectively have been provided.

CONCLUSIONS

It appears that a lot of research is being undertaken to innovate OPAC and make it a sophisticated and efficient retrieval tool in accessing vast library resources. This development is of course through different generations. In the process, attempts are being made to incorporate newer and newer features in OPAC, such as non-conventional bibliographical details viz. table of contents, summary and physical features of book in order to add enhanced browsable graphical aspects to it. It is found that these features try to reduce the cognitive load as experienced by the users while handling analytical search strategies on OPAC. Presenting a simulated view of the library layout is a further step in this direction.

However, not much work has been done in the area of the actual location of library material and its retrieval through OPAC, in spite of its perceived need by library users and researchers alike. In a large library location of library material is significant from the users' point of view. Such libraries have large location data. No independent system has yet been identified, capable of handling such data. The classification scheme is used with too many restrictions to adequately perform subject grouping, and location retrieving functions. As such both the functions are not fully justified. The location data is usually handled arbitrarily by this technique. From the readers' point of view, in addition to retrieve the classification details and location statements, there is need to have actual location finding mechanism. This requires a system which can handle the geographical location aspects of material across large library premises. Such a system must be capable of identifying and tracking down huge location data of library material independent of classification numbers.

The present study attempts to provide solutions to these problems. On one hand the study has taken initiative to discuss computerised modelling, the shelving process of the library by introducing idea of "sheiving model", which can monitor the shelving activity of the entire library collection and give guidance to library staff. On the other hand it has theoretically proved how Geographical Information Systems (GIS) can successfully be used to handle the location data and provide location finding tool in OPAC itself. This is done by introducing the new concept of mapping of library collection using GIS. Such mapping is possible because most libraries virtually have an immobile shelf arrangement as argued in the study.

While considering immobile library shelving situation as the basis behind the idea of mapping of library collection the static library environment or the storehouse situation has been used as the starting point as one of the developmental stages in presenting the further stage.

Mapping of the library collection using GIS makes it possible to see not only a simulated view of the library building and the broad location of the material but it makes if possible to see the location of library material specifically upto a rack and shelf level in a digitized floor map alongwith its bibliographical and other details through OPAC.

After discussing the new concept theoretically in detail, wherein the study has suggested the idea of mapping library collection as a solution to the location problem in the library, the next natural question was to look into the feasibility of the idea. The study has, therefore, attempted to go one step ahead and has demonstrated the idea through the development of a prototype. For this purpose a suitable library where such application would be useful was identified. Floor diagrams with geographical locational features incorporated in them were prepared and digitized. To facilitate this a suitable GIS package which could serve the purpose was identified. In order to use it for library applications certain modifications were required, which were achieved programatically. Further, an experimental attribute database was created. The map data thus created through digitization was linked to attribute data. The system was carefully run to get the required results. The analysis and observations made on the results led to the conclusion that such an application is feasible and recommended for implementation.

In addition to important uses of GIS in library discussed above, the study concentrated on other two applications of GIS viz. (1) use of GIS in providing information services based on spatially referenced data and (2) use of GIS in networking/mapping of library resources.

Regarding the first application it is stated that presently such application is limited only to census related information and there is a need to widen these types of services in other areas and look into systematically establishing such services in the library. The study has, therefore, discussed most of the aspects of establishing of such services in the library and of the extension of these services in other areas. It has invited attention towards the future potential of such services and the responsible role of librarians, and towards the need for the creation of spatially referenced databases by the library.

In respect of the second application the study has explored the use of GIS in mapping/networking of library resources. It has presented various advantages of using GIS for this purpose. The concept of thematic mapping of library resources and using internet for the purpose suggested are going to be useful in solving some of the major problems in library networking which are faced presently.

The study has further attempted to explore various uses of GIS in other areas of library management. Such as (a) use of GIS in mapping geographic facets against subject content by exploiting geographic treatment given by Colon and other classification schemes; (b) use of GIS in mapping booksellers and distributors; (c) use of GIS in user education; and (d) use of GIS in user study. While identifying all these probable uses of GIS in the library, the study has attempted to bring a new geographic dimension in all these areas of library management which was not conceptualized so far. The advantages drawn in this respect lead to the conclusion that such applications are going to provide further innovative ways for the organization of libraries.

Although it was not planned as it was difficult to demonstrate all the above referred uses of GIS in library due to time and financial limitations and other constraints, exception was made in case of "mapping of library collection" and it was decided to show it experimentally. Judging by the success of the first application which is demonstrated it is concluded that all the other uses of GIS as discussed in the present study are feasible and would be useful.

RESEARCH FINDINGS

The study has explored various uses of GIS in the library as was envisaged in the introductory chapter and the discussion has concluded that all those uses are feasible. It was, however, observed that in spite of having enough potential for these applications these issues have not attracted enough attention of the researchers, resulting in a scarcity of literature in this area.

Secondly while working on the topic especially on the development of the prototype it was found that :

(i) It was difficult to demonstrate the concept of "mapping of library collection", through a fully developed system. The difficulty was in terms of required financial resources, highly technical expertise and the hardware-software support. Such support was also not possible from the university under whose auspices the research work was conducted nor was it possible through the organisation with which the researcher is associated. Otherwise, it would have been extremely useful if such system could have been fully developed, put into use and the user response to the same observed and analyzed. Here the necessity of the establishment of such a research project of a non-commercial nature was in mind.

(ii) It was also observed that the GIS package available was not suitable for this type of applications. It was necessary to put in a good amount of programing efforts to prepare the prototype. However, looking into development of GIS technology it is felt that still more convenient and a higher level GIS could be available in future at less cost wherein such application could easily fit-in without involving so much programing and other efforts. Moreover, it could give better results than were presented in the present prototype, which was completed under too many constraints and with whatever support was available to the researcher.

RECOMMENDATIONS

It may not be possible to accept and implement the ideas and concepts discussed in the present study immediately as most of them are very new to the field of library science or even to the field of GIS. As concluded in this study all the uses and applications suggested are likely to add new (geographical) dimensions to the existing library systems especially to OPAC retrieval. Secondly, due to the probable involvement of other research issues (as listed below) in the study it is recommended to have wider discussion on the topic in various forums of libraries and librarians at national and other levels.

Following are the issues which may emerge out of the present study which seek further research :

(i) Using depth classification and observing its advantages in browsing.

(ii) Use of multimedia and animation technology in showing graphical features of library material through OPAC.
(iii) Sharing high level GIS by the group of libraries in providing GIS based library systems through network facilities.
(iv) Use of GIS and remote sensing technique in monitoring movement of material in the library.
(v) Exploiting spatial facet of classification.

It is quite possible that the thinking in the direction of the present study has already begun. However, such scarce attempts need to be reflected in the library literature, or need to be translated into English if already published in non-English languages. To obviate this, it is recommended to establish a special research group to discuss the topic "Use of GIS in Library" on the lines of discussion presented in the book.

APPENDICES

APPENDIX - I

LIST OF GIS VENDORS, THEIR PRODUCTS AND OTHER USEFUL INFORMATION

(A) Contact Details for the three Major GIS Software Workstation Vendors

- ARC/INFO
  ESRI System Inc.
  380, New York Street
  Redlands, CA 92373
  714/793-2853

  Products : Arc/Info
  Arc/View
- ERDAS Inc
  2801, Buford Highway, NE
  Atlanta. GA 30329
  404/248 - 9000
- Integraph Corporation
  One Madison Industrial Park
  Huntsville AL 35807
  800/826-3515

(B) Contact and Product Details of IBM Compatible GIS Softwares

- Caliper Corp.
  1172, Beacon St.
  Newton. MA 02161
  617/527-4700

  Product : GIS Plus $ 2995.

- Map Info Corp.
  200 Broadway
  Troy. N. Y. 12180
  800/327-8627

  Product : Map Info $ 995
  Map Info for Windows $ 995
  Macintosh Product Map Info $ 995
- Strategic Mapping Inc.
  4030 Moor Parts Ave
  Suit 250
  Sun Jose CA 95117
  408/985-7400

  Products : Atlas GIS $ 2,495
  Atlas Pro (DOS) $ 795
  Atlas Mapmarker for Windows $ 495
  Import-Export $ 495
  Atlas Stript $ 495
  Macintosh Product Atlas Promac $ 975
  Atlas Mapmarker $ 495

  Atlas GIS Requires following hardware
  IBM XT
  640 KRAM
  EGA Monitor
  10 Megabytes of Hard Disk
  MS DOS 3.0 or later
- Tactics Intl.
  16 Haver Hill St.
  Third Floor
  Andorer. MA 01810
  800/927-7666

  Product : Tactician $ 2,495
  Macintosh : Tactician $ 2,495

  Hardware requirement for Tactician
  Macintosh II
  4 Megabytes of RAM
  Colour Monitor
  20 MB of Hard Disk
  System 6.0.5

(C) Contact and Product Details of Macintosh Compatible GIS Softwares

- PC Globe Inc.
  4700, Smith McDintock
  Tempe A2 85282
  800/336-6314

  Product : Descarters $ 395
- Geo Query Corp.
  475 Alexis & Shuman Blvd.
  Suit 380 E
  Naperville, IL, 60563
  800/541 0181.
  Product GeoQuery - $ 395
- John Wiley & Sons Inc.
  605, Third Avenue
  New York, NY 10158.
  Product : Map II $ 99
- Map 'n' Fact is available from
  Broderbuw Software Inc.
  500, Redwood Boulevow P.O. Box 6121.
  Novato, CA 94948
- Global Explorer Map GO and Map Expert are
  available from Delome Mapping
  P.O. Box 298, Freeport ME 04032
- ARL Electronic Atlas is accessible on the
  www on home page
  Http:// www-Map.Lib. Umn, edu/arl. html

Some Useful Addresses and Other Information

- Environmental Systems Research Institute Inc.
  380, New York Street,
  Redlands, CA 92373-8100
- ERDAS
  2801, Buford Highway, NE
  Suite 300 Atlanta
  GA 30329-2137.

- Geographic Resources Analysis Support Systems (GRASS)
  U.S. Army Construction Engineering Research, Laboratory,
  P.O. Box. 4005 Champaign
  IL 61824-43005
- Sammamish Data Systems
  1813, 130th Avenue NE, Suit 216,
  Belevue, WA 98005-2240
- Data User Services Division, US Bureau of the Census, Washington DC 20233-8360
- NCGIA : National Centre for Geographic Information and Analysis
  NCGIA Homepage address is
  http:www. ncgia. ucsb. edu.
- ARC View is one product in a line of geographic software and data from the Environmental System Research Institute (ESRI). Their major product is ARC INFO runs on mainframe is a fully analytic command driven vector based geographic information system. Arc View may not have the power of Arc Info but does include a graphical user interface. While a GIS may require month or training for proficiency casual users may be able to master mapping programs like Arc View in a few hours. In one sense Arc View is the front end through which Arc Info and other data can be read.
- The Consortium for Intl. Earth Science Information Network (CIESIN) is developing and maintaining an internet directory to spatial data and metadata. The CIESIN home page http:// www ciesin. orgm.
- ARC/INFO^{R} and ARC/VIEW^{R} are the registered trademarks of Environmental System Research Institute.
- ESRI's recent entry ARC View 2.0 is an almost complete revision from the first release. It is vastly improved tool box help system ability to create and work with non-topological shape files, advanced table management including statistical analysis support for multimedia charting and graphing and customizing with the Avenue application development-language make it very strong product. Arc View allows the user to save map displays in project components called views which do not actually contain the data but which acts as pointers to them. The view preserves all the symbology table of contents (list of themes) and user behind features of the

original display. Another strength of Arc View is its ability to use direct files created by Arc/Info, ESRI's widely distributed flagship GIS product. The price of Arc View 2.0 in a special package of data for publicly accessible libraries and K 12 group is $ 495. The avenue language is an additional $495 and (list prices - $ 1995 for Arc View with Avenue, and $995 without it). Arc/View is positively reviewed by cart-write who call it as an "Extraordinary product".

The hardware configuration, cost of packages and also product manufactures and their addresses are based on information sources dated in 1995-96 and they need to be confirmed against current details as GIS technology is continuously updated. It was felt that as one may not get sorted details of GIS in library application point of view. One can use the above information while planning such applications.

Researcher acknowledges the following resources used to prepare the above appendix

- Carl Franklin, "Introduction to Geographical Information Systems : Linking Maps to Databases, *Database* 15, No. 2 (April 92) : 12-21.
- Karl Longstreth, "GIS Collection Development, Starting and Training", *Journal of Academic Librarianship"* 21, No. 4 (July 1995): 265-274.
- Brent Allison, University of Minnesota : Remote Access and the Internet, *Journal of Academic Librarianship* 21, No.4 (July 1995) : 283-287.
- Patrick McGlomery and Melisa Lamount, "New Opportunities and Challenges : Geographical Information Systems in Libraries", *Database* (December 1994) : 34-44.
- Therela C. Straller, "Desktop GIS in Libraries, Technology and Costs : A View from the New York State", *Journal of Academic Librarianship* 21. No.4 (July 1995) : 278-282.

APPENDIX-II

SCREEN PHOTOGRAPHS (PLATES) SHOWING OUTPUT OF PROTOTYPE RESULTS

This comprises of the results, in terms of screen photographs of the prototype developed to demonstrate the concept of "Mapping of Library Collection and examples of searching books with their geographical locations along with bibliographic details.

PROTO-TYPE RESULTS

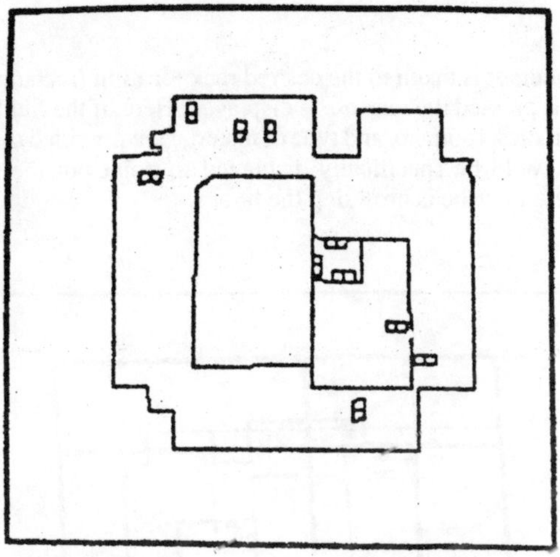

Enter mode S. Search by location I. Search by book details (S/I)

Plate 1 : This is the first screen of prototype "GIS based Library Information System". There are two way searches available (1) S=search by location i.e., from spatial to non-spatial and (2) I=search by book details i.e., from non-spatial data to spatial locations. Same floor diagram but in a concised form has been used for the demonstration. However, a very few rack location places as seen in the diagram along with their attribute data have been used while experimenting and demonstrating the concept of "Mapping of Library Collection" and finding location of books in digitised floor diagram.

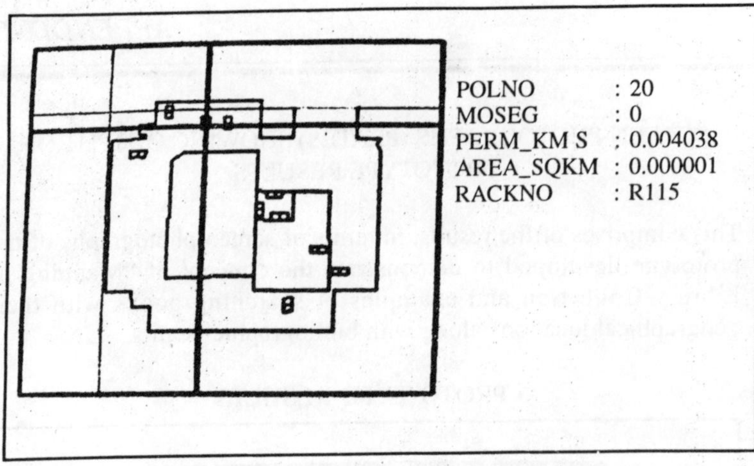

Plate 2 : The cursor is taken to the desired rack location (rectangle). After enter key is pressed the rack no. is displayed. Here, if the fields like location/floor/stack room no. and type of material are provided then those details also could be specifically displayed to point out the reader at what location he/she is browsing the books.

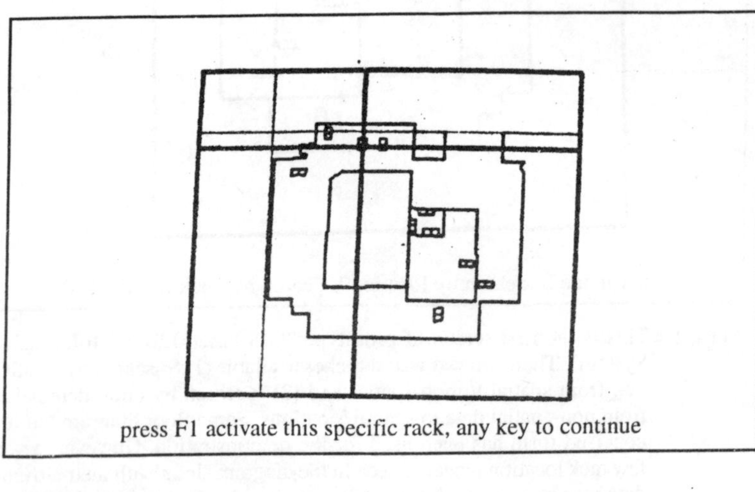

Plate 3 : For further activating the particular rack, F1 key is pressed.

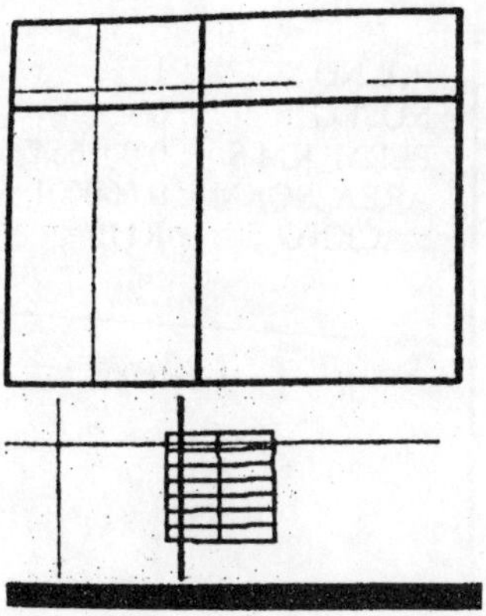

RACKNO	: R115
SHELFNO	: 1
ACCNO	: 170055
TITLE	: Intl. jr. of geographical informa
AUTHOR1	:
AUTHOR 2	:
SUBJECT	: GIS
PUBLISHER	: Taylor and Francis
YEAR	: 1987
CLASSNO	: 05:911:681.3
AUTHMARK	:
TYPE	: Journals

Plate 4 : After activating the particular rack in terms of shelves, the system takes the browser to the first shelf of the rack and displays the details of first book as arranged in the particular sequence i.e. either class numberwise or titlewise. Here onwards one could browse details of all the books available on that shelf in the sequence. Similarly one can switchover from shelf to shelf to see the details of books kept there. Thus in spatial to non-spatial browsing would be possible anywhere in the floor map.

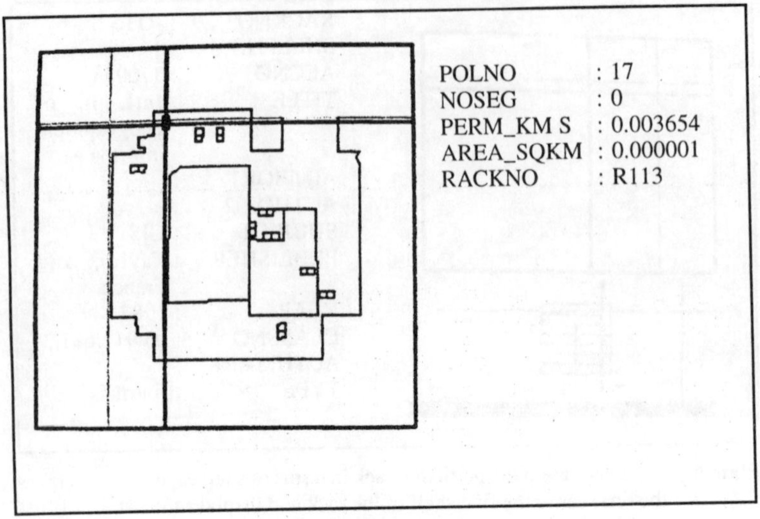

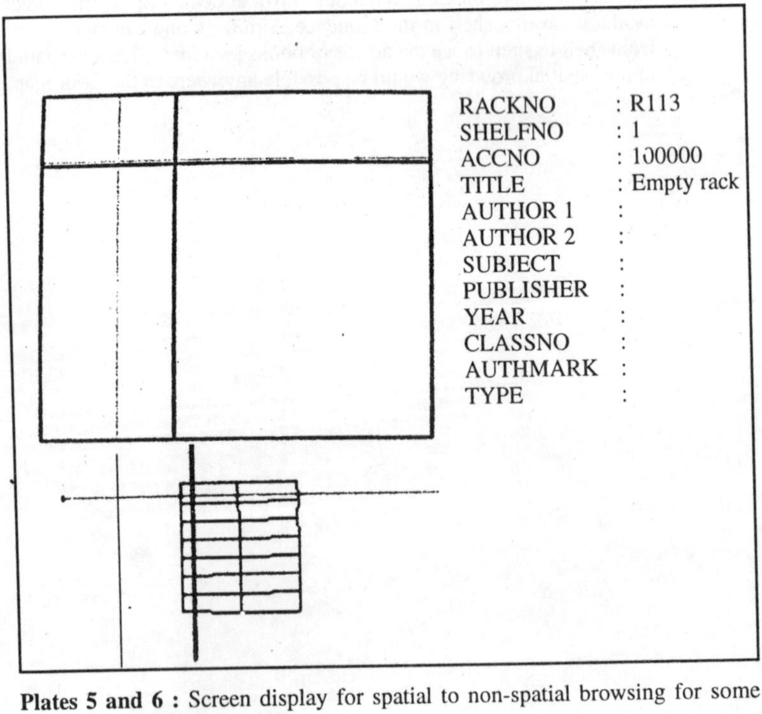

Plates 5 and 6 : Screen display for spatial to non-spatial browsing for some other rack

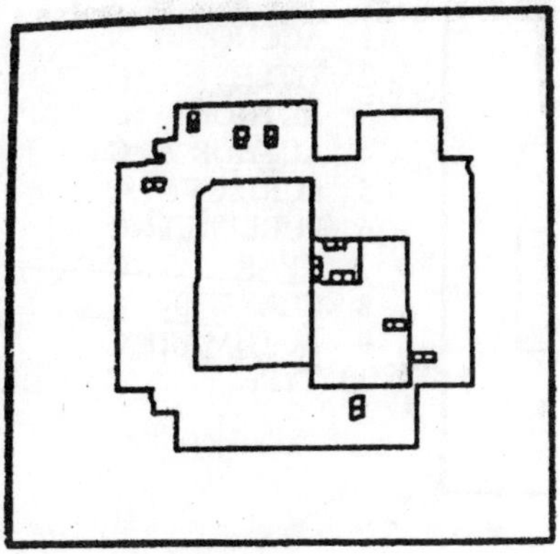

Press <Esc> to select

1 ACCNO N
2 TITLE C
3 AUTHOR1 C
4 AUTHOR 2 C
5 SUBJECT C
6 PUBLISHER C
7 YEAR C
8 CLASSNO C
9 AUTHMARK C
10 TYPE C

ATTRIBUTE DATABASE
FILES STRUCTURE

Plate 7 : Here the next mode has been selected for non-spatial to spatial browsing i.e. this screen mode is accessed by pressing "I" key to search by book details. The set of fields displayed on the right side indicate that one can choose any one of those fields for searching books with that particular field details in mind. <Esc> key pressed to select the field.

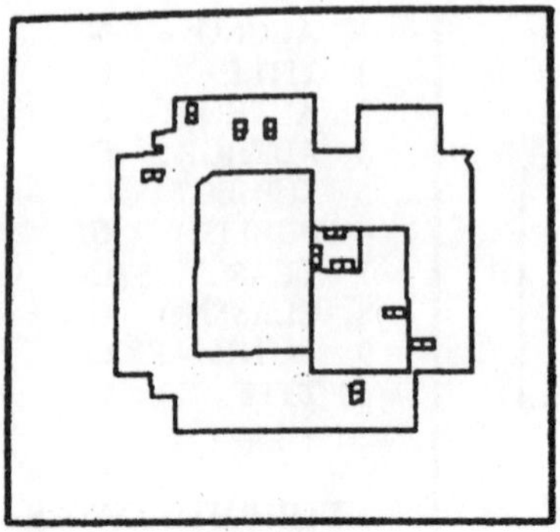

1 ACCNO N
2 TITLE C
3 AUTHOR1 C
4 AUTHOR 2 C
5 SUBJECT C
6 PUBLISHER C
7 YEAR C
8 CLASSNO C
9 AUTHMARK C
10 TYPE C

Give the field no. : 5

Plate 8 : Here field number 5 is given to select search books by the field "subject" or "Key-word."

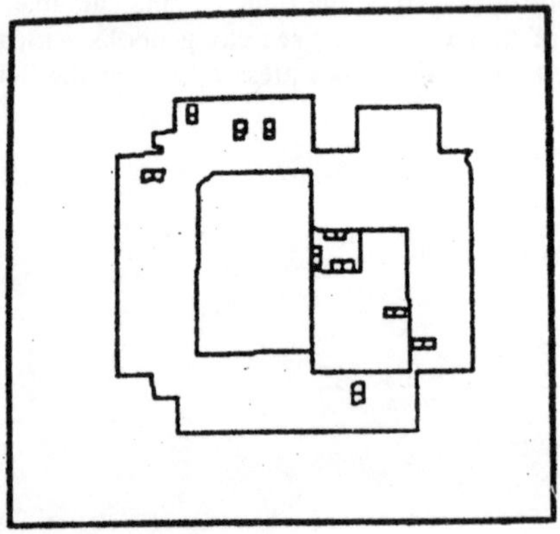

1 ACCNO N
2 TITLE C
3 AUTHOR1 C
4 AUTHOR 2 C
5 SUBJECT C
6 PUBLISHER C
7 YEAR C
8 CLASSNO C
9 AUTHMARK C
10 TYPE C

Give the field value : GIS

Plate 9 : Here the value of field no. 5 is given as "GIS" i.e. to search availability of books in the library on Geographical Information Systems (GIS).

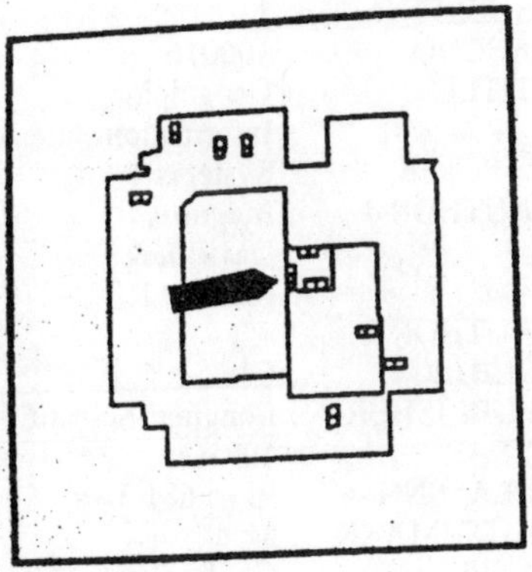

RACKNO	: R101
SHELFNO	: 1
ACCNO	: 100015
TITLE	: International GIS World Source
AUTHOR 1	:
AUTHOR 2	:
SUBJECT	: GIS
PUBLISHER	: GIS World Inc
YEAR	: 1996C
CLASSNO	: 911 : 681.3
AUTHMARK	: int
TYPE	: Ref. Books

10 entries are selected, use ->, <- to browse data, <Esc> to stop

Plate 10 : After pressing the enter key the search result is displayed at the bottom as"10 entries are selected" which means the library has 10 books having "GIS" as subject. The details of the first book out of those ten are displayed at right hand side at the same time the location of book i.e. rack R101 where the book has been kept is also highlighted with red colour. After seeing this first book the browser can look for other entries out of those ten books by using right arrow and left arrow keys.

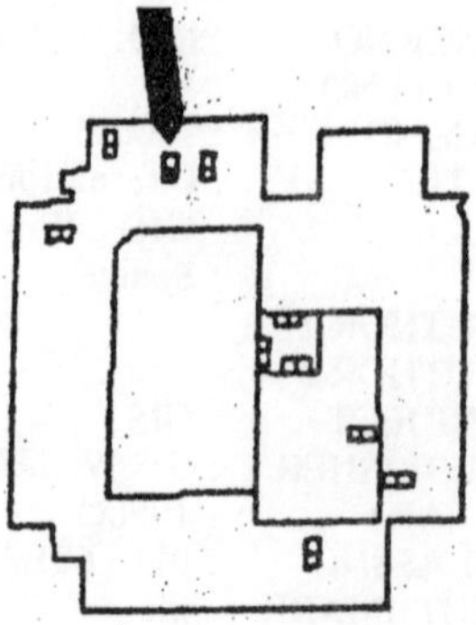

RACKNO	: R116
SHELFNO	: 1
ACCNO	: 100016
TITLE	: Geographic Information Systems
AUTHOR 1	: Maguire, David J. & Micheal R.
AUTHOR 2	:
SUBJECT	: GIS
PUBLISHER	: Longman Scientific
YEAR	: 1991C
CLASSNO	: 911 : 681.3
AUTHMARK	: Mag
TYPE	: GN Books

10 entries are selected, use ->, <- to browse data, <Esc> to stop

Plate 11 : In the same group of ten books with subject as "GIS" the second entry is displayed by pressing right arrow key. After pressing the right arrow key details of another books are displayed at right hand side and at the same time its location to the particular rack is correspondingly highlighted in red colour.

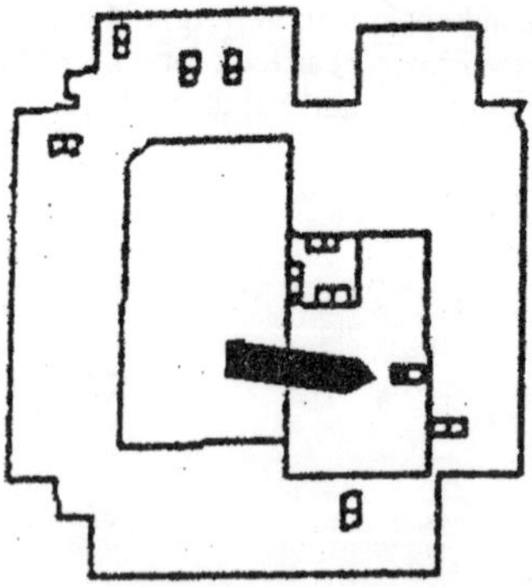

RACKNO	: R107
SHELFNO	: 1
ACCNO	: 100017
TITLE	: Principles of Geographic information
AUTHOR 1	: Burrough, P.A.
AUTHOR 2	:
SUBJECT	: GIS
PUBLISHER	: Oxford University Pr
YEAR	: 1986C
CLASSNO	: 911 : 681.3
AUTHMARK	: Bur
TYPE	: GM Books

10 entries are selected, use ->, <- to browse data, <Esc> to stop

Plate 12 : Similarly the third entry in the same sequence is displayed with its respective location highlighted.

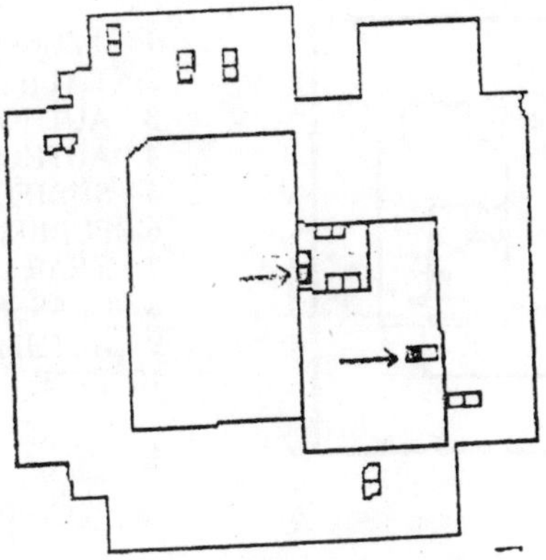

	Reference Books	*GN Books*	*Journals*
RACKNO	: R101	R107	R115
SHELFNO	: 1	1	1
ACCNO	: 100015	100017	170055
TITLE	: International GIS World Source Book	Principles of Geographic Infor Systems	Intl. Journal of GIS
AUTHOR 1	:	Burrough, P.A.	
AUTHOR 2	:		
SUBJECT	: GIS	GIS	GIS
PUBLISHER	: GIS World Inc.	Oxford Univ. Pr.	Taylor & Francis.
YEAR	: 1996	1986	1987
CLASSNO	: 911 : 681.3 (083)	911:681.3	05:911:681.3
AUTHMARK	: Int	Bur	
TYPE	: Ref. Books	GN Books	Journals

Plate 13

Note : Ideal software would display locations of all the ten entries found in single floor diagram or across the different floor diagrams, group them depending on type/form of material by using different colour graphics and would allow browser to see the bibliographic details one by one as per choice. This could give an idea exactly how library collection can be mapped using GIS. It could be on similar lines as shown in plate 13.

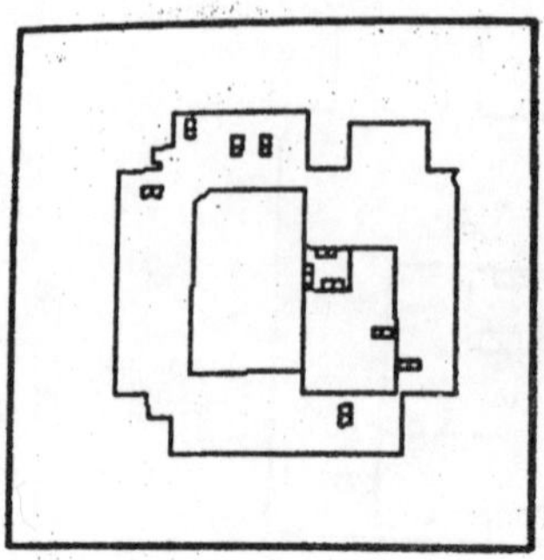

1 ACCNO N
2 TITLE C
3 AUTHOR1 C
4 AUTHOR 2 C
5 SUBJECT C
6 PUBLISHER C
7 YEAR C
8 CLASSNO C
9 AUTHMARK C
10 TYPE C

Give the field no. : 2

Plate 14 : In the second example of searching location by book details with field No. 2 i.e. "title" is chosen.

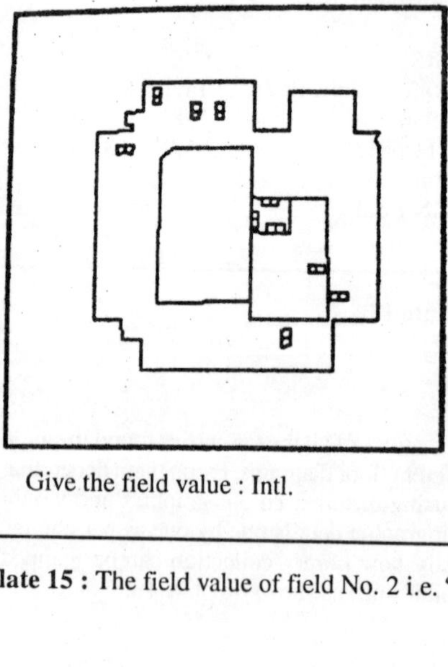

1 ACCNO N
2 TITLE C
3 AUTHOR1 C
4 AUTHOR 2 C
5 SUBJECT C
6 PUBLISHER C
7 YEAR C
8 CLASSNO C
9 AUTHMARK C
10 TYPE C

Give the field value : Intl.

Plate 15 : The field value of field No. 2 i.e. "title" is given as "Intl."

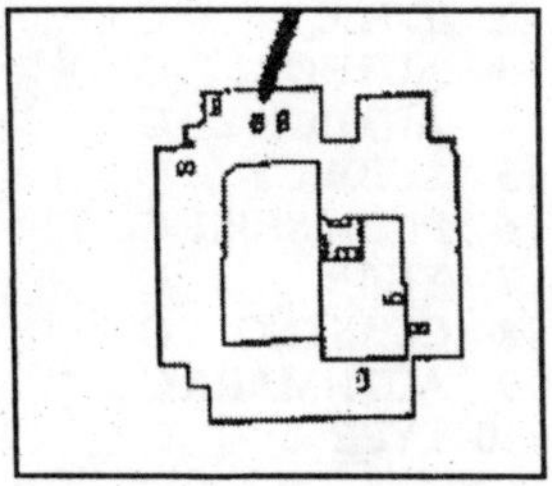

RACKNO : R115
SHELFNO : 1
ACCNO : 170055
TITLE : Intl. Jr. of geographical inform
AUTHOR 1 :
AUTHOR 2 :
SUBJECT : GIS
PUBLISHER : Tailor & Francis
YEAR : 1987C
CLASSNO : 05:911:681.3
AUTHMARK :
TYPE : Journals

3 entries are selected, use ->, <- to browse data, <Esc> to stop

Plate 16 : After pressing the enter key the search result is displayed at the bottom as "3 entries are selected" and at the same time for the first one of those 3 books, details are displayed at right hand side and its location in the floor diagram that is rack position is shown as highlighted in red colour.

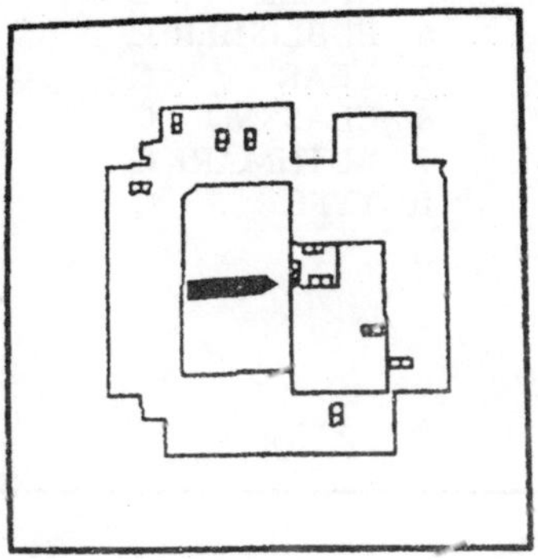

RACKNO : R101
SHELFNO : 1
ACCNO : 100015
TITLE : International GIS World Source
AUTHOR 1 :
AUTHOR 2 :
SUBJECT : GIS
PUBLISHER : GIS World Inc.
YEAR : 1996C
CLASSNO : 911:681.3
AUTHMARK : Int
TYPE : Ref. Books

3 entries are selected, use ->, <- to browse data, <Esc> to stop

Plate 17 : Shows the details of second book in the same sequence with its location in the floor diagram.

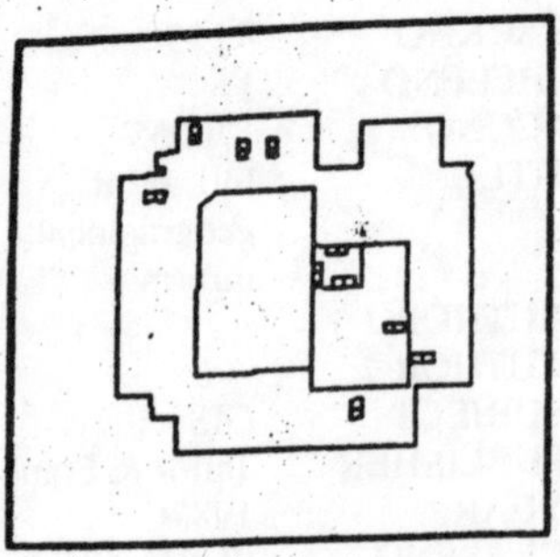

1	ACCNO	N
2	TITLE	C
3	AUTHOR1	C
4	AUTHOR 2	C
5	SUBJECT	C
6	PUBLISHER	C
7	YEAR	C
8	CLASSNO	C
9	AUTHMARK	C
10	TYPE	C

Give the field no. : 3

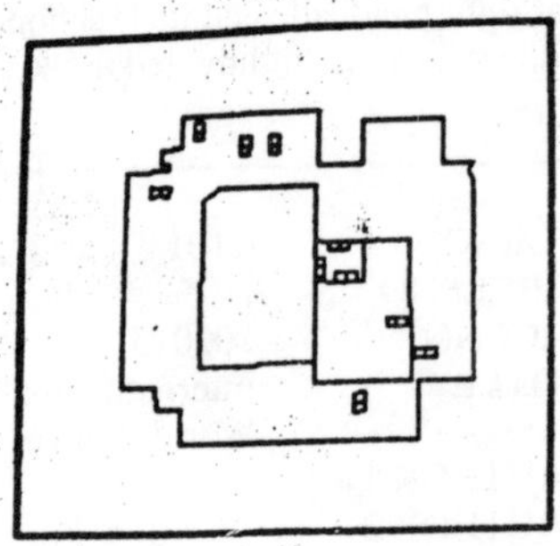

1	ACCNO	N
2	TITLE	C
3	AUTHOR1	C
4	AUTHOR 2	C
5	SUBJECT	C
6	PUBLISHER	C
7	YEAR	C
8	CLASSNO	C
9	AUTHMARK	C
10	TYPE	C

Give the field value : Ranganathan

Plates 18 & 19 : Here author field no. 3 i.e. "Author 1" is chosen for search. The field value of the author field is given as "Ranganathan".

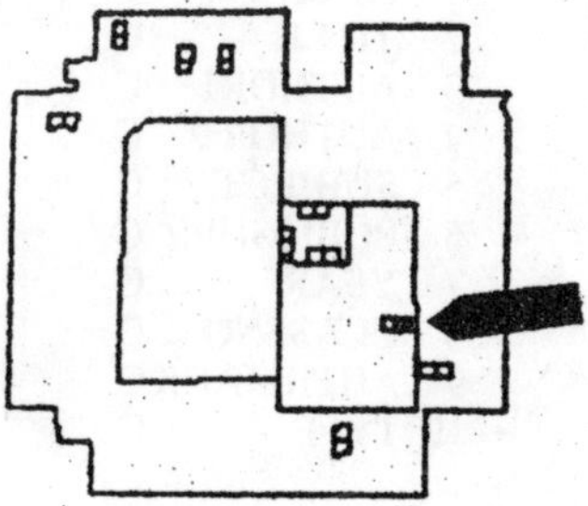

RACKNO : R100
SHELFNO : 1
ACCNO : 150046
TITLE : Documentation and its facets
AUTHOR 1 : Ranganathan, S.R.
AUTHOR 2 :
SUBJECT : Libscience
PUBLISHER : Asia Publishing
YEAR : 1963C
CLASSNO : 002
AUTHMARK : Ran
TYPE : GN Books

2 entries are selected, use ->, <- to browse data, <Esc> to stop

Plate 20 : After enter key is pressed results are displayed as—"2 entries are selected." Here it means that there are two books with author as "Ranganathan." At the same time for first of those two books the bibliographic details as well as location mark are displayed simultaneously.

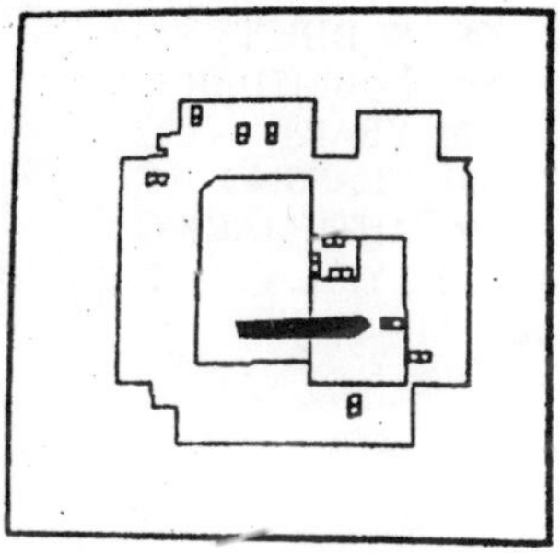

RACKNO : R107
SHELFNO : 6
ACCNO : 150045
TITLE : Colon Classification. Edition
AUTHOR 1 : Ranganathan, S.R
AUTHOR 2 :
SUBJECT : LIBSCIENCE
PUBLISHER : Sarda Ranganathan
YEAR : 1987C
CLASSNO : 025.49
AUTHMARK : Ran
TYPE : GN Books

2 entries are selected, use ->, <- to browse data, <Esc> to stop

Plate 21 : The details of second book with author as "Ranganathan" are displayed with location.

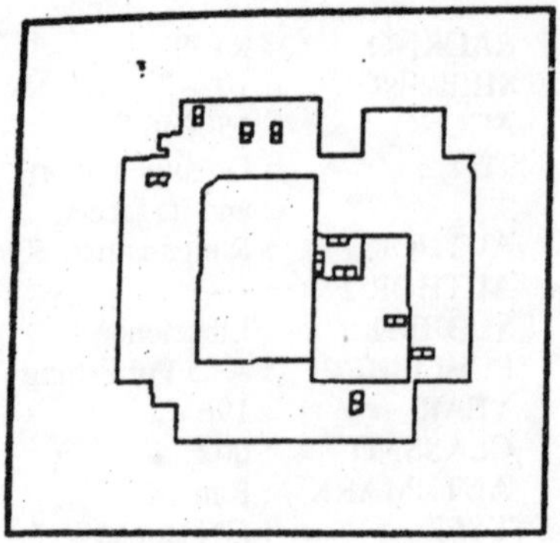

1	ACCNO	N
2	TITLE	C
3	AUTHOR1	C
4	AUTHOR 2	C
5	SUBJECT	C
6	PUBLISHER	C
7	YEAR	C
8	CLASSNO	C
9	AUTHMARK	C
10	TYPE	C

Give the field no. : 8

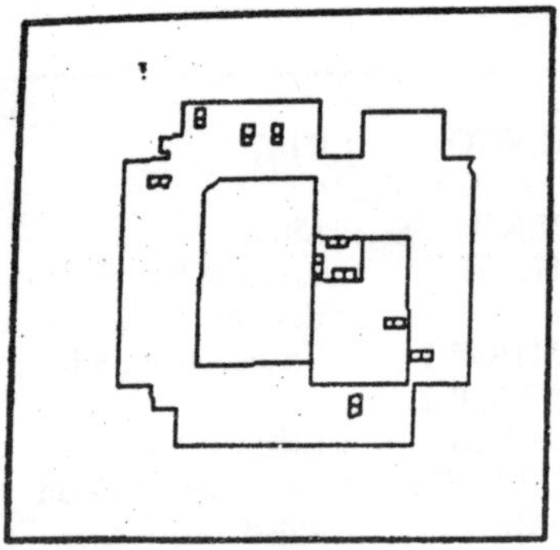

1	ACCNO	N
2	TITLE	C
3	AUTHOR1	C
4	AUTHOR 2	C
5	SUBJECT	C
6	PUBLISHER	C
7	YEAR	C
8	CLASSNO	C
9	AUTHMARK	C
10	TYPE	C

Give the field value : 517

Plates 22 & 23 : Here some other field was chosen i.e. No. 8 indicating "class No." The subsequent screen the value of the class No. is given as "517" which refers to "Mathematical Analysis".

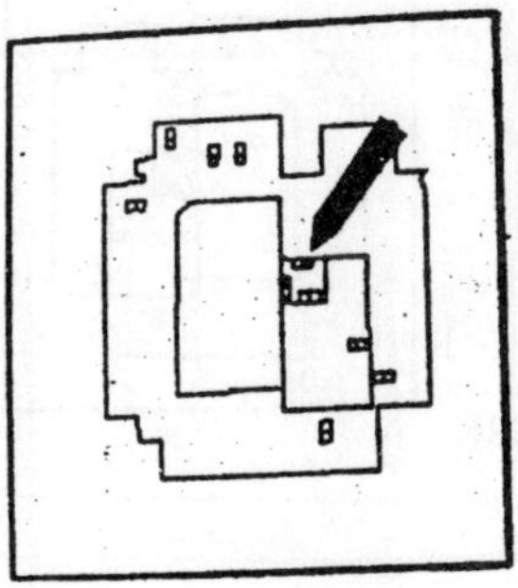

RACKNO	: R104
SHELFNO	: 2
ACCNO	: 11002
TITLE	: Notes on complex function theo
AUTHOR 1	: Sarason, Donald
AUTHOR 2	:
SUBJECT	: Mathematics
PUBLISHER	: Hindustan Books
YEAR	: 1994C
CLASSNO	: 517
AUTHMARK	: Sar
TYPE	: GN Books

1 entry is selected, use ->, <- to browse data, <Esc> to stop

Plate 24 : After pressing the enter key the result of the search has been displayed as "1 entry is selected." At the same time it is seen that the bibliographical details have been given in right hand side and the location of that particular book also has been shown in the library floor map in the respective rack in red colour.

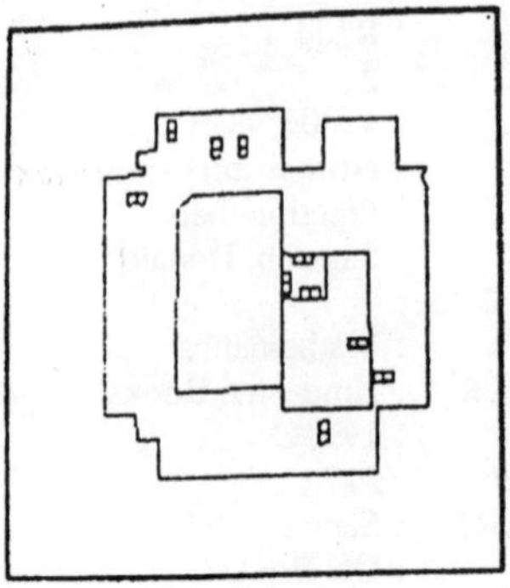

1 ACCNO
2 TITLE
3 AUTHOR1
4 AUTHOR 2
5 SUBJECT
6 PUBLISHER
7 YEAR
8 CLASSNO
9 AUTHMARK
10 TYPE

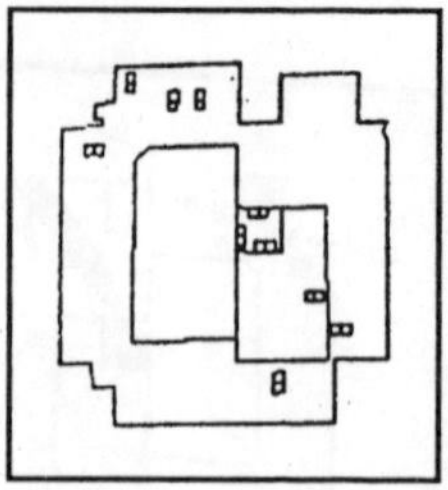

Give the field no : 3

Give the field value : Maguire

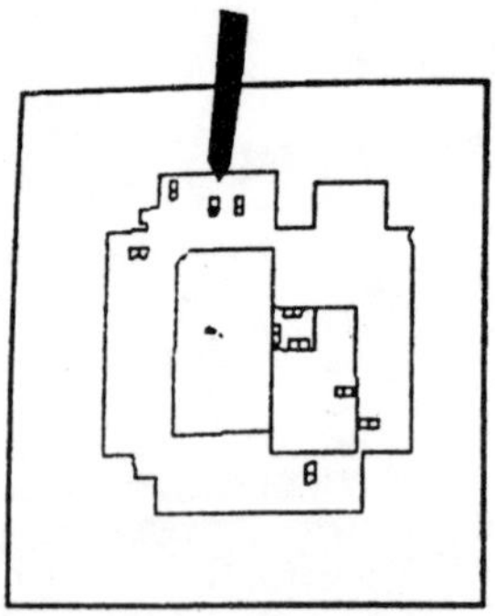

RACKNO	: R116
SHELFNO	: 1
ACCNO	: 100016
TITLE	: Geographic Information Systems
AUTHOR 1	: Maguire, David J. & Micheal R.
AUTHOR 2	:
SUBJECT	: GIS
PUBLISHER	: Longman Scientific
YEAR	: 1991C
CLASSNO	: 911:681.3
AUTHMARK	: Mag
TYPE	: GN Books

1 entry is selected, use ->, <- to browse data, <Esc> to stop

Plates 25-27 : This search provides another example of search by author i.e., field No. 3 where the field value given as "Maguire". The result indicates that there was one entry and for the same bibliographical details have been shown at right hand side and the location of that book has also been indicated by highlighting the respective rack in the red colour.

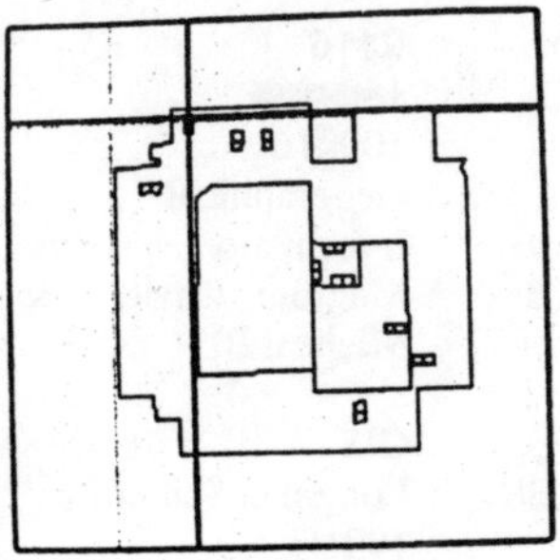

POLNO : 16
NOSEG : 0
PERM_KM S : 0.004101
AREA_SQKM : 0.000001
RACKNO : R112

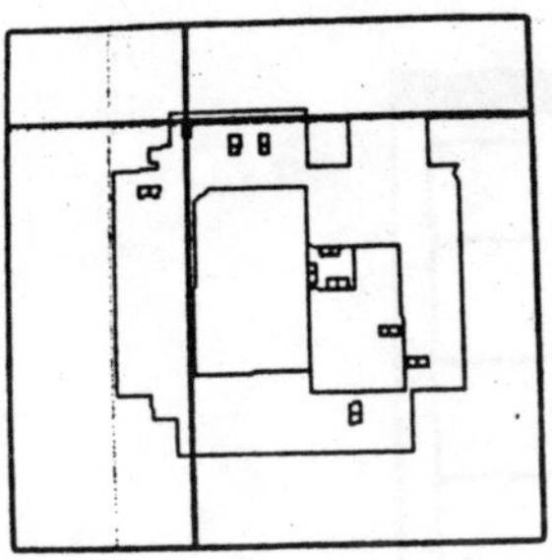

POLNO : 10
NOSEG : 0
PERM_KM S : 0.004053
AREA_SQKM : 0.000001
RACKNO : R103

Plates 28-29 : These plates show examples of spatial to non-spatial searching for three different racks

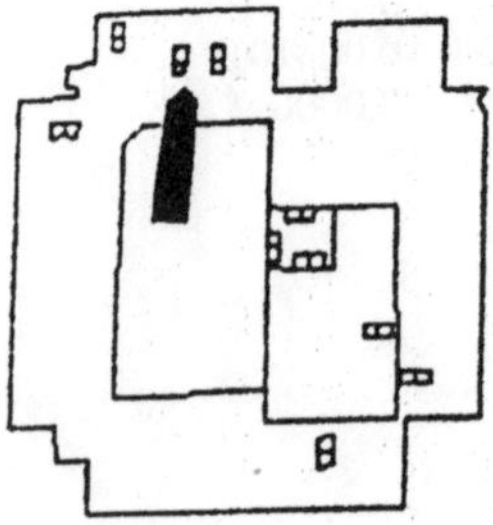

RACKNO	: R116
SHELFNO	: 1
ACCNO	: 100016
TITLE	: Geographical Information Systems
AUTHOR 1	: Maguire, David J. & Micheal R.
AUTHOR 2	:
SUBJECT	: GIS
PUBLISHER	: Longman Scientific
YEAR	: 1991C
CLASSNO	: 911 : 681.3
AUTHMARK	: Mag
TYPE	: GN Books

10 entries are selected, use ->, <- to browse data, <Esc> to stop

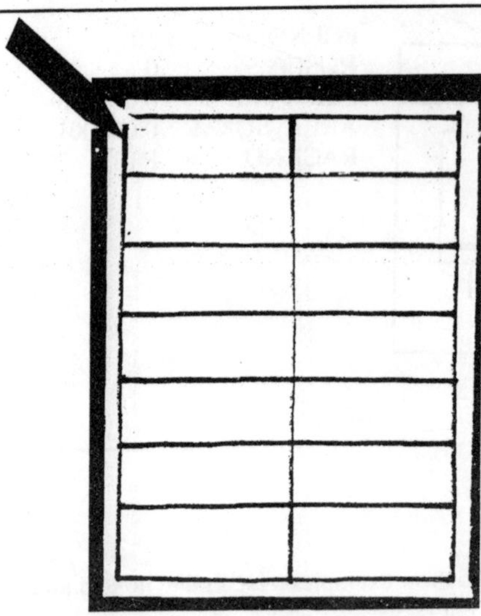

Plate 30 : This shows how the non-spatial to spatial search could go one step ahead and show the highlighted particular shelf of the rack where that book has been shelved. This is logical next step. As seen in the plate No. 4 it is possible to activate the particular shelf. However, this step is not covered in the prototype. What is shown is the logical example of the same.

Bibliography*

Abbott, Lisa T. and Carolyn D. Argentati., "GIS A New Component of Public Services". *Journal of Academic Librarianship* 21, No. 4 (July 1995) : 251-256.

Adler, Prudence. "Special Issue of Geographic Information Systems (GISs) and Academic Libraries : An Introduction". *Journal of Academic Librarianship* 21, No. 4 (July 1995) : 333-335.

Aiten G.D., G.V.O. Malley and H. E. MacFarlane. "Maps and Tomorrows "Libraries : Map Producers View". *Globe* 39 (1993) : 33-39.

Allen, R.S. "An overview of Federal Geographic Data Committee National Spatial Data Infrastructure. National Spatial Data clearing house and Digital Geospatial Metadata Standard what will it mean for tomorrow's libraries"?. *Geography and Map Division Bulletin* No. 180 (June 1995) 2-28.

Allison, Brent, "University of Minnesota : Remote Access and the Internet". *Journal of Academic Librarianship* 21, No. 4 (July 1995) 283-287.

Andrew M. Hawkins, (1994), Geographical Information System (GIS): Their Use as Decision Support Tools in Public Libraries and the Integration of GIS with Other Computer Technology, *New Library World,* 95 (7).

Antenucci, John C. *et al. Geographic Information Systems : A Guide to Technology.* London : Chapman Hall, 1991.

Aplin, G. "Changing Maps of Changing Cities : Mapping Urban Change". *Globe*, No. 41 (1995) : 19-28.

Arms, W.Y. "Keynote address : The Virtual Library". *Proceedings of*

* The bibliography has been prepared as per style of Chicago Manual of Style, 14th Edn., New Delhi : Prentice Hall of India, 1996.

the UK Office for Library Networking Conference. April 2-5. 1993. London : Meckler, 1993.

Arnfield, R. "Electronic Mail : A Road to the Information Super Highways". *Library Journal* 119, No. 13 (August 1994): 16-19.

Arnoff, Stanley, *Geographic Information Systems : A Management Perspective,* Ottawa : Ontario Canada, WDL Publications; 1989 : 1.

Austhof, B. "Global Explorer 1.0". *Technical Services Quarterly* 12, No. 3 (1995) : 65-69.

Australian Institute of Cartographers Mapping Service 1994 Conference, Gold Coast, 4-8 September". *Globe* No. 41 (1995) : 105-106.

Ayyar, M. "Mapping with internet : Transmitting maps electronically". *Globe*, No. 41 (1995) : 88-94.

Beheshti, Jamshid, Valerie Large and Mary Bialek. "PACE: A browsable graphical interface". *Information Technology and Librarians* 15, No. 4 (December 1996) : 231-240.

Bell, S.J. and M. Halperin. "Testing the realibility of cellular online searching". *Online*, 19, No. 5 (September/ October 1995) : 15-24.

Blair, N.L. and C. J. Manson. "Need for a national geologic mapping index databases", Paper presented at the 28th Meeting of the Geoscience Information Society, ed. by Connie Wick, Boston, Massachusetts, 25-28, October 1993 Alexandria, Virginia : Geoscience Information Society, 1994 : 179-195.

Bob Batchelder. "Natural resource mapping in Calgary : An overview". *Western Association of Map Libraries Information Bulletin*, 13, No. 2 (March 1982) : 130-138.

Boisse, Joseph A. and Mary Larsgaard. "GIS in Academic Libraries: A Managerial Perspective". *Journal of Academic Librarianship* 21, No. 4 (July 1995) : 288-291.

Bookscope (CD-ROM). Wood Bridge : Primary Source Media, 1996.

Braken, Ian. *A Classification of Geographic Information Systems Literature and Applications.* Wales and South West Regional Laboratory : University of Wales College of Cardiff, 1989.

Brook, Peter Smith. "Geographic Information Systems is this the end of the map?". *Globe*, No. 27 (1987) : 48-70.

Burrough, P.A. *Principles of Geographic Information Systems for Land Resources Assessment.* Oxford : Oxford University Press, 1986.

Campbell, Harold G. "Geographic Information Systems : Education for non-computer oriented college students". *SIGGSE Bulletin* 26, No. 3 (September 1994) : 11-14.

Campbell, T. "Possibilities for the retroconverted map files". *IFLA Journal* 20, No. 1 (1994) : 46-52.

Carrington, David K. "Information and recent developments of interest to users of MARC Map". *Inspel*, 14, No. 3/4 (1979) : 62-66.

Cartwright, W. "New medium and presentation devices : Implications for cartographers and map collections". *Globe*, No. 39 (1993) : 18-25.

Case, Donald Owen. "Conceptual Organisation and retrieval of text by historians : The role of memory and metaphor". *Journal of American Society for Information Science* 42, No. 9 (October 1991) : 657-668.

Cataloguing and Classification Quarterly 13, No. 3 & 4 (1991) 1-243 (*Special issue : Enhancing Access to information Defining Catalogs for the 21st Century.*

Catenazzi, N. and L. Sommaruga. "Hyper Library : a formal model for an electronic library based on hyperbooks". *Journal of Documentation* 5, No. 3 (September 1995) 244-270.

Cheviric, Joan F. "Getting started : ready set get organised". *Journal of Academic Librarianship* 21, No. 4 (July 1995) : 292-296.

Chiang, K.S. and D.Y. Rieger. "Introduction to extracting and mapping census data". *Reference Services Review* 23, No. 1 (1995) ; 29 38.

Chiang, Katherine S. "Geographical Information System : Services at Mann Library". (Sidebar 3) in Susan J. Barnes. "The electronic library and public services". *Library Hi Tech* 12, No. 3 (1994) : 53

Citchfield, Charles A., and Murilyn L. Norstedt. "Coded holdings : A primer for new users". *Serials Review* 14, Nos. 1-2 (1988) : 81-88.

Clark, P.M. "Geographic analysis of downloaded data", Paper presented at National Online Meeting 1995, ed. by Marth E. Williams, Learned Information Inc; New York, 2-4 May 1995 Medtow, New Jersey : Learned Information Inc., 1995 : 57-64.

Clarke, Keith C. "Geographic Information Systems Definitions and Prospects". *Geographic and Map Division Bulletin* No. 142 (December 1985) : 12-18.

Clementini, E. and P.D. Felice. "Object oriented modelling of geographic data". *Journal of the American Society for Information Science* 45, No. 9 (October 1994) : 694-704.

Clifford, A. Shaffer. "Data Representations for Geographic Information Systems". *Annual Review of Information Science and Technology.* 27 (1992) : 135-172.

Cline, N.M. and P.S. Adler. "GIS and research libraries : one perspective". *Information Technology and Libraries* 14, No. 2 (June 1995) : 111-115.

Cobb, David A. "Developing GIS relationship". *Journal of Academic Librarianship* 21, No. 4 (July 1995) : 275-277.

Cogalton, Russell G. and Kass, Green. "The Abcs of GIS : An Introduction to Geographic Information Systems" in *Wetland and Environmental Applications or GIS,* Ed. by John B. Lyon and McCarthy. New York : CRC Press Inc., 1995 : 9.

Cohn, A.G., *et al.* "A logical approach to representing and reasoning about space". *Artificial Intelligence Review* 9, Nos. 4-5 (October 1995) : 255-259.

Cooper, J. "Digital Map data and Geographical Information Systems". *Highways and Transportation* 42, No. 2 (February 1995) : 14-15.

Cowen, David J. "Importance of GIS for the average persons". GIS in Government of the First Federal Perspective" *Proceedings of the First Federal Grographic Technology Conference*, Fort Williams, 10 : GIS World Inc. 1994.

Cox, Allen B. "An overview to Geographic Information Systems". *Journal of Academic Librarianship* 2, No. 4 (July 1995) : 237-249.

Cox, C. and G. Fitzpatrick., "Exploring the future of map libraries". *LC Information Bulletin* 53, No. 7 (April 1994): 142-143.

Crockford, N. "A plain guide to space planning". *ASLIB Information* 21, No. 1 (January 1993) : 24-25.

Dangermond, J. "Trends in geographic information systems software". *Proceedings IGIS : The Research Agenda.* Washington D.C. Nasa, 1987.

David Green, Terry Bossomaier (2003). Online GIS and Spatial Metadata, *Electronic Library*, 21 (3).

de Gruyter, M.L. "Texas State Electronic Library". *Texas Library Journal* 71, No. 1 (Spring 1995) : 24-28.

Deo Data Institute. GIS Awareness. *Information Technology Training*

Initiative. Geo Data Institute, University of Southampton, Southampton, UK 1992.

Duckett, B. "Mute testimony : Yorkshire's historic library buildings". *Library Review* 44, No. 3 (1995) : 44-68.

Dwyer, Catherine G., Eleano A. Gossen and Lynne M. Martin. "Known item search failure in OPAC". *RO* 31, No. 2 (Winter 1991) : 228-236.

Eaton, Gale. "Lost in the library : are spatial skills important in shelf searching". *Journal of Youth Services in Libraries* 5, No. 1 (Fall 1991) : 77-86.

Eaton, J. "1991 census on CD-ROM". *Managing Information* 1, No. 11/12 (November-December 1994) : 55-56.

"Electronic mapping software can aid business." *Information Today* 13, No. 2 (February 1996) : 24.

Environmental Systems Research Institute. Arc View 2.0d for windows. The Geographic Information System for Everyone, Redlands, Calci : ESRI : 1995 (ISBN 1-879102-44-7) (Reviewed by R. Lorson, in *Library Quarterly* 66, No. 3 (July 1996) : 313.

Eva, Margaret. "Geology and maps : Notes on study town". *Globe*, No. 23 (1985) : 64-70.

Fitzpartrick, Gary L. "Indirect access to maps via online databases". *Proceedings of 4th National Online Meeting*. New York, April 12-14, 1993 : 139-146.

Fleet, C. "Comparing Automated Map Cataloguing Systems . Pilot study based in the National Library of Walse". *Program* 28, No. 3 (July 1994) : 223-237.

Fleming, Myrna. "Cataloguing of Cartographic Materials". *Western Association of Map Libraries Information Bulletin* 11, No. 3 (June 1980) : 176-181.

Foster, G. "Mapping the Library Landscape". *Library Review* 44, No. 3 (1995) : 5-7.

Fouty, G. "Journal Finder : Making the link between journal abbreviations and stack locations". *Library Software Review* 11, No. 3 (May-June 1992) : 13-16.

Frank, S. "Cataloguing digital geographic data in the information infrastructure : A literature and technology review". *Information Processing Management* 30, No. 5 (1995) : 587-606.

Franklin, Carl. "An introduction to Geographic Information Systems :

Linking Maps to databases". *Database* 15, No. 2 (April 1992) : 17-21.

Franklin, Carl. "An introduction to Geographic Information Sytems : Linking Maps to databases". *Database* (April 1992) : 12-15 and 17-21.

Gambosi, G. "Basic research actions for geographic object oriented databases systems", Universita di Roma "La Sapienza" Dipartimento di Informatica Sisemistica Ustizio Ragioneria - Sett Entrate, Viale Manzoni 30, 00185 Rome, ITALY.

Garnsworthy, J. and G. Tail. "Do you know where I could find out about (Government Spatial Data)". *Burisa*, No. 114 (July 1994) : 11-13.

Gluck, M. "Prospecting for users based GIS Process", *Proceedings of 35th Annual Meeting of American Society for Information Science,* Pittsburgh, 26-29 October 1992, Ed. by Bebra Shaw, Medford New Jersey : Learned Information Inc for American Society for Information Science 1992 : 242-248.

Golden, Gary A. "Taming the unfriendly system : Microcomputer al patron terminal to access on online catalog" in *What is your friendly*, ed. by F.W. Lancaster. Urbuna : University of Illinois, 1987 : 69-79.

Goodchild, M.F. and K.K. Kemp eds. "Introduction to GIS". *NCGIA core curriculum.* Santa Barbara, California : University of California, 1990.

Gopen, D. Kaye. "The Virtual Library, Knowledge Society and the Librarian" in *Virtual Library : Visions and Realities* ed. by Lauerna M. Saunders. West Port : Mockler, 1993.

Griffith, J. and K. Bonair. "Satellite remote sensing and information management". *FID News Bulletin* 45, No. 3 (March 1995) : 81-84.

Grimshaw, David J. "Geographic Information Systems in Business". *Encyclopedia of Library and Information Science* 54 (1994) : 187-202.

Grivel, L., P. Mutschke and Polanco, "Thematic mapping on bibliographic databases by cluster analysis : a description of SDOC environment with SOLIS". *Knowledge Organisation* 22, No. 2 (1995) : 70-77.

Gruse, Larry. "Cartography's photographic revolution Microcartography". *Wilson Library Bulletin* No. 60 (October 1985): 17-20.

Guptill, S.C. "Spatial data standards and information policy". *Government Information Quarterly* 11, No. 4 (1994) : 387-401.

Harrington, T.P. "Reference Assistant : A Hyper Text touch screen program to assist users of a learning resources centre". *Journal of Educational Media and Library Science* 30, No. 3 (Spring 1993) : 221-230.

Hayes, S. and D. Brown. "The Library as a business : Mapping the pervasiveness of Financial relationship in today's library". *Library Trends* 42, No. 2 (Winter 1994) : 404-419.

Haynes, D. "Business process reengineering and information audits". *Managing Information* 2, No. 6 (June 1995) : 30-32.

Herkovitz, Sara *et al.* "Development of spatial structure of libraries : The case as Tel Aviv". Libri 41, No. 2 (June 1991) : 121-131.

Hernandez, T. *et al.* "GIS and retail location decision making : Survey results". *Burisa* No. 121 (December 1995) : 2-5.

Hildreth, Charles R. "Advancing towards the E^3 OPAC : the imperative and the path" in *Think tank on the present and the future of the online catalogue :* Proceedings occasional paper No. 9. Chicago American Library Association on Reference and Adult Service Division, 1991 : 5-19.

Hildreth, Charles R. *Online Public Access Catalogue : The User Interface.* Ohid : OCLC, 1982.

Hill, Janet Swan. "Developments in map cataloguing at the Library of Congress". *Special Library* 68, No. 4 (April 1977) : 149-154.

Hochsmann, Dieter and Schlitt Gerhand. "Spatial Situation and Planning of buildings for academic libraries in the new federal States". *Bibliotheksdienst* 25, No. 12 (1991) : 1881-1896.

Hudson, Alice C. "Conversion to automated cataloguing at map division NYPL". *Special Libraries* 7, No. 2 (February 1976) : 97-101.

Hunt, Li and Marc Joselyn. "Maximizing Accessibility to spatially Referenced Digital Data". *Journal of Academic Librarianship* 21, No. 4 (July 1995) : 257-265.

Huxhold, W.B. and A.G. LKevinsohn. 'Managing Geographic Information System Projects". *Canadian Journal of Information and Library Science* 20, No. 2 (July 1995) : 42-43.

Impact GIS Project Near Completion Information". *Management Report* (January 1996) : 6-7.

International GIS World Source Book (Annual). St. Collins Co : GIS World Inc, 1989).

Isaac, P. "British book trade index: Development and progress". *Library Review* 44, No. 3 (1995) : 17-23.

Jacob, M.E.L. "32nd Annual UIUC Clinic an attendee's view". *Library Hi-Tech News.* No. 126 (October 1995) : 10-11.

James, Robert. "Libraries in mind : How can we see users perspective of libraries". *Journal of Librarianship* 15, No. 1 (January 1983) : 19-23.

James, S. "Mapping the library landscape". *Library Review* 44, No. 3 (1995) : 4-71.

Jansen, D. "Preserving evidential value in geographic information systems". *Archives and Museum Informatics* 8, No. 3 (1994) : 192-204.

Jeanne Boston, Charles W. Dean, Hugh Phillips, Nolan F. Pope (1998) The Public Electronic Library : Integrating GIS resources and tools, *Library Hi Tech,* 16 (3).

Jingfeng Xia (2004) GIS in the management of library pick-up books, *Library Hi Tech,* 22 (2).

John Tombarge (1999) Using management software for geographic information systems (GIS), *Bottom Line,* 12 (4).

John, S.A. "Data integration in a GIS : The question of data quality". *ASLIB Proceedings* 45, No. 4 (April 1993) : 109-119.

Kane, William P. "A Hyper card call number directory : Using stacks to find stacks". *College and Research Library News* 50, No. 7 (July/August 1989) : 576-577.

Karrow, Robert W. "Innocent Pleasures ISBD (CM) AAICR and Map Cataloguing Part-I". *International Cataloguing* 12, No. 1 (January/ March 1983) : 10-12.

Karrow, Robert W. Innocent Pleasures. ISBD (CM) AACR and Map Cataloguing". *Geography and Map Division Bulletin* No. 126 (December 1981) : 2-12.

Karrow, Robert W. "Multidimensional Cataloguing rules and IFLA recommendations : 3 innocent pleasures ISBD (CM) AACR II and Map Cataloguing Part 2". *International Cataloguing* 12, No. 2 (April-June 1983) : 19-21.

Kevin D. Davie, (2001) The ARL Geographic Information Systems Literacy Project, *Library Hi Tech,* 19(1).

Kimball, Glenn S. Color Fundamental Part two : "Color scanning and data compression". *Document Image Automation* 11 No. 3 (May-June 1991) 156-162 (Part one published in Document Image Automation 11, No. 2 (March-April 1991) : 74-81.

Lai, Pohchin and Charles F. Gillies. "The impact of geographic information systems on the role of spatial data libraries". *International Journal of Geographic Information Systems* 5, No. 2 (1991) : 241-251.

Laing, Ken and David Mander. "Where history and technology meet". *The Library Association Record* 98, No. 8 (August 1996) : 420-427.

Lambert, Steve. *CD-ROM : the new papyrus*, Redmond Wash: Microsoft, 1986.

Lancaster, F.W. *et al.*, "The relation between literature scatter and journal accessibility in an academic special library". *Collection Building* 11, No. 1 (1991) : 19-21.

Lang, Laura. "Mapping the Future of Map Librarianship". *American Librarian* 28, No. 10 (November 1992) : 880-883.

Larkin, V. "Tiger : The computerization of the art of mapping". *New Jersey Libraries* 26, No. 4 (February 1993) . 28-30.

Larsgaard, M.L. and L. Carver :"Accessing Spatial data online : Project Alexandrian". *Information Technology and Libraries* 14, No. 2 (June 1995) : 93-97.

Laurini, R. and D. Thompson. *Fundamantals of Spatial Information Systems.* New York : Academic Press, 1992.

Lee, Newton S. "Multimedia Visualizer : An animated objective based OPAC". *Information Technology and Libraries* 10 (December 1991) : 297-310.

Liisa, L. and M. Talikka. "The use of the Virginia Tech Library Systems (VTLS) Facilitates the location of grey literature". *IATUL Proceedings* 2 (1993) : 162-168.

Line, Maurice B. *et al.* "How golden is your retriever : Thoughts on library classification". in *Lines of Thought : Selected papers of Maurice B. Line*, ed. by. L.J. Anthony. London : Clive Bingley, 1988.

Litchfield, Charles A. and Marilyn L. Norstedt. "Coded holdings : A primer for new users". *Serials Review* 14, Nos. 1-2 (1988) : 81-88.

Longstreth, Karl. "GIS : Collection development staffing and training". *Journal of Academic Librarianship* 21, No. 4 (July 1995) : 267-274.

Loosee, R.M. "The relative shelf location of circulated books : A study of classification users and browsing". *Library Resources and Technical Services* 37, No. 2 (April 1993): 197-209.

Lukes, David. "Computer Mapping of Local Authorities". *Burisa* No. 72 (February 1986) : 2-3.

Maguire, David J., Michael F. Goodchild and David W. Rhind. eds. *Geographical Information Systems; Principles and Application,* 2 vols. Harlaw : Logman Scientific and Technical, 1991.

Making GIS a part of Library Services" (Special Issue on GIS). *Information Technology and Libraries* 14, No. 4 (July 1995).

Manchester, P. "Mapping out the market". *PC Week* (12th February 1994) : 20-22.

Mandernack, S.B. "Street atlas U.S., V. 2.0". *Geography and Map Division Bulletin.* No. 181 (October 1995) . 48-51.

Mansfield, G. "Mapping communication networks in the directory". *Computer Networks and ISDN Systems* 26, No. 3 (November 1993) : 327-335.

Manth, R. "Find it on the map". *Byte* 20, No. 6 (June 1995) : 9-14.

Mapping the market : A survey of co-operative developments in the field of business information provision". *British Library Research and Development Department BLRD Reports,* No. 6148 (1994) : 32.

Marek, K. "Union Catalog on Compact disk : A cooperative pilot project". *Nebraska Library Association Quarterly* 26, No. 2 (Summer 1995) : 51-53.

Martie Pienaar, Pieter Van Brakel (1999) The changing face of geographic information on the web: A breakthrough in spatial data sharing, *Electronic Library,* 17 (6).

Martin, Wayne. "Color micrographics and mapping today". *Geography and Map Division Bulletin,* No. 152 (June 1988) : 35-38.

Martyn, J. "Geological Mapping on Macintosh". *Globe,* No. 41 (1995) : 27-31.

McGlamery, Patrick and Medisa Lamont. "New Opportunities and Challenges : Geographic Information Systems in Libraries". *Database* (December 1994) : 35-44.

Mellendorf, S.A. "Pounding the pavement with purpose : Utilizing the information superhighways for daily work tasks". *RO* 35, No. 2 (Winter 1995) : 231-235.

Minton, James O. "OCLC map cataloguing : An emerging standard practice". *Geography Map Division Bulletin,* No. 13 (September 1978) : 25-40.

Mondschein, L.G. "The use of environmental spatial information of Johnson and Johnson", Paper presented at the 14th National Online Meeting 1993 ed. by Martha E. Williams, Learned Information INC., New York, 4-6 May 1993. Medford, New Jersey : Learned Information Inc, 1993 : 319-329.

Mooney, Des. "Map Information, a valuable state resource". *Globe,* No. 35 (1991) : 9-11.

Moore, Patrica A. "Topographic maps in US Libraries". *International Library Review* 19, No. 3 (July 1987) : 201-223.

Morris, B.A. *et al.* "From map catalogue to databases with graphical in view". *Liber Bulletin,* No. 28 (1986) : 22-29.

Morris, Barbara. "Carto-Net : A Cartographic Information Retrieval System". *British Library,* 1987-88. (Research Paper No. 33).

Morris, Barbara. "Carto-Net : Graphic retrieval and management in an automated map library". *Geography and Map Division Bulletin* No. 152 (June 1988) : 19-35.

Morris, Barbara. "Carto-Net : Graphic retrieval and management in an automated map library". *Inspecl* 22, No. 2 (1988) : 91-112.

Morris, Barbara. "Carto-Net : Networking map cataloguing records". *British Library* 1988 (British Library Research Paper No. 53).

Multidimensional mapping of libraries based on share holdings in the OCLC Online Union Catalogue". *OCLC*; 1983.

Murr, L.E., J.B. Williams and R.E. Miller. "Mapping Information Delivery Networks : The objectives, the methods, the benefits and the model". *Library Hi Tech Issue* 13, Vol. 4, No. 1 (Spring 1986) : 51-63.

National Seminar on Geographic Information Systems (GIS) for development planning, 12-14 February 1997. Madras. Madras : The Ranganathan Centre for Information Studies, 1997.

Newby, Gregory B. "Virtual Reality". *Annual Review of Information Science and Technology* 28 (1993) : 187-229.

North, Gory W. "Will your library be the spatial data information of the future". *Inspel* 23, No. 2 (1989) . 130-138.

Noyons, E.C.M. and A.F.J.V. Rann. "Bibliometric Cartography of Scientific and technological developments of an R&D Field". *Scientometrics* 30, No. 1 (May 1994) : 157-173.

Obokaha, N.P. and Arokoyu S.B. "The influence of geographical location on Public Library use : A case study from the developing country". *Geography and Map Division Bulletin,* No. 163 (March 1991) : 30-42.

OCLC. "Building a catalogue of internet resources". *Advanced Technology Libraries* (October 1994). OCLC

OCLC. "Netfirst a comprehensive database of internet accessible resources". *Advanced Technology Libraries* (May 1995).

OGRIP Advisory Committee's First Year Report, Department of Administrative Services, State of Ohio, 1996.

Opadeyi, J.A. "Planning and implementation strategies for the acquisition and utilization of GIS". in *Proceedings of the 4th Annual Technical Conference of ARETT,* November 1990. Trinidad.

Opadeyi, Jacob. "Geographic Information Systems : An Information Management Tool". *FID News Bulletin* 45, No. 3 (March 1995) : 77-80.

Parkar, J.R. "Cartographic Communication : What does the future hold". *Globe,* No. 35 (1991) : 4-8.

Parkins, C.R. "Leave it to the labs? Options for the future of map and spatial data collections". *European Research Libraries Cooperation* 5, No. 3 (1995) : 312-329.

Parkins, C.R. and R.B. Parry. "Changing patterns of map availability". *Inspecl* 22, No. 1 (1988): 56-59.

Parry, R.B. "The electronic map library : New maps, new uses and new users". *European Research Library Cooperation* 5, No. 3 (1995) : 262-273.

Parsons, K.A. "Software review : Map expert with street Atlas USA". *Technical Services Quarterly* 12, No. 2 (1994) : 69-75.

Patrick McGlamery (2000) Lifting the fog: maps and spatial information emerging as grey literature, International Journal on Grey Literature, 1 (1).

Piquettee, Constance M. "Map Librarianship : A Selective and annotated bibliography". *Western Association of Map Libraries Information Bulletin* 15, No. 1 (November 1983) . 67-73.

Poulter, Alan. "Building a browsable virtual reality library". *ASLIB Proceedings* 46, No. 6 (June 1994) : 151-155.

Poulter, Alan. "Towards a virtual reality library" *ASLIB Proceedings* 45, No. 1 (January 1993) : 11-17.

Price, G. "Mapping the future". *Planning Week* 2, No. 13 (31st March 1994) : Suppt xi-xii.

Pritchard, J.A.T. "Geographic Information Systems : A report from GIS 95". *Information Management Report* (November 1995) : 17-19.

Ranganathan, S. R. *Five Laws of Library Science.* Bombay : Asia Publishing House, 1963.

Rao, Mukund, V. Jayarama and M.F. Chandrashekhar. "Organising Spatial Information Systems around a GIS core (Database Organisation and integrated analysis using GIS)". Special Pub No. ISRO NNRMS SP 70-94 Bangalore : Indian Space Research Organisation, December 1994.

Raper, J.F. and N.P.A. Green. "Development of Hypertext Based Tutor for Geographic Information Systems". *Brit Journal of Educational Technology.* 9 (1989) : 3-23.

Renand, R. "Building digital library". College and Research Library News 56, No. 8 (September 1995) : 547-548.

Reynolds, J. "Geographical Information for Retail Research Constraints and Opportunities". *Burisa* No. 112 (February 1994) : 2-4.

Rieger, O.Y. "Introducing numeric CD-ROMS library". *Micro-Computers for Information Management* 10, No. 2 (June 1993) : 93-118.

Riston, Walter W. "The greening of map librarianship". *Geography Map Division Bulletin,* No. 111 (March 1978) : 2-9.

Ronald C. Jantz (1997). Geographical information systems at Princeton University : Evaluation and exploration of a new library service, *Collection Building,* 16 (3).

Rowley, J.E. and L.G. Peacefull. "A Review of the Geographic Information Systems market place in the United Kingdom". *Information Services and Use* 15, No. 2 (1995) : 103-115.

Rowley, Jennifer. "Nature and applications of Geographic Information Systems". *ASLIB Managing Information* (April 1994) : 1-4.

Rucker, Newton W. "Defense mapping agency libraries". *Science and Technology Libraries* 5, No. 3 (Spring 1985) : 39-44.

Salzberg, B. "On indexing spatial and temporal data". *Information Systems* 19, No. 6 (September 1994) : 447-465.

Sargent, K.A. and Jonathan C. Matti. "The national geographic mapping program : A revitalisation of geologic mapping in the United States. Collection for the future archivists, curators, historians, bibliographers, sperl", Paper presented at 22nd Meeting of Geo-Science Information Society. *Phoenix*, Arizona 26-29 December 1987 (18) 91-101.

Schroeder, J.R. "AACR II Map Cataloguing Allert". *Geography and Map Division Bulletin*, No. 114 (December 1987) . 29.

Schroeder, J.R. "Perspectives on map cataloguing and classification". *Library Trends* 29, No. 3 (Winter 1981) : 419-438.

"Setting Standards for geographic data". *Burisa*, No. 115 (September 1994) : 12-16.

Small, R.V. and S.M. Ferreira. "Information location and use, motivation and learning patterns which using print or multimedia information resources". *Journal of Educational Multimedia and Hypermedia* 3, No. 314 (1994) : 251-273.

Smith, L.C. *et al.* "32nd Annual UIUC clinic highlights spatial and cartographic information technology". *Library Hi-Tech News*, No. 126 (October 1995) : 1-9, 21.

Solimine, G. "Spatial Organisation within the library system". *Biblioteche Oggi* 11, No. 8 (September 1993) : 42-45.

Spiess, E. "Some problems with the use of electronic atlases". *European Research Libraries Co-operation* 5, No. 3 (1995) : 235-244.

Star, J. and J. Estes. *Geographic Information Systems : An introduction.* New Jersey : Prentice Hall, 1989.

Stephens, Irving E. "Getting more out of call number : Displaying holdings, locations and circulation status". *Cataloguing and Classification Quarterly* 13 No. 3/4 (1991) : 97-102.

Stevens, Alan R. "National Cartographic Information Centre : An information resource on mapping products for the nation". *Science and Technology Libraries,* No. 3 (Spring 1985) : 25-38.

Strasser, Theresa C. "Desktop GIS in Libraries : Technology and Costs : A view from New York State". *Journal of Academic Librarianship* 21, No. 4 (July 1995) : 287-282.

Studwell, W.E. and E.K. Rast. "Format integration and spatial data : A preliminary view". Western Association of Map Libraries Information Bulletin 24, No. 3 (July 1993) : 186-188.

Su, Shiao Feng. "Dialog with an OPAC : How visionary was Swanson in 1964?". *Library Quarterly* 63, No. 2 (1994), 130-161.

Swanson, Don R. "Dialog with a catalog". *Library Quarterly* 34 (January 1964) : 113-125.

Swanson, Don R. "Requirements study for Future Catalogs". *Library Quarterly* 42 (July 1972) : 302-15.

Theresa C. Strasser (1998). Geographic information systems and the New York State Library : Mapping new pathways for library service, *Library Hi Tech*, 16(3).

Triangle Research Libraries Networking Governing Board, "Concept Paper : The Future Online Systems at TRLN". November 27, 1990.

Tusa, B.M. "Overview of application of automation to special collection, maps and archives". *Information Technology and Libraries* 12, No. 4 (December 1993) : 405-411.

Tyacke, Sarah. "Digital maps and map libraries". *Inspecl* 21, No. 1 (1987) : 48-54.

Tyckoson, David A. "The 98 per cent solution: The failure of the catalog and role of electronic databases". Technicalities 9, No. 2 (February 1989) : 8-12.

University Grants Committee. "Capital provision for University Libraries". (Report of the Working Chairman Richard Atkinson). London : HMSO, 1976.

University Grants Committee. Committee on Libraries Report No. 1967, pp. 213-228. Appendix 5, Report by Brimingham University Library on surveys carried out in 1964 on the use of Libraries of undergraduates, graduate students and staff. London : HMSO, 1967.

Vassilakopoulos, M. and Y. Manolopouloss. "Analytical comparison of spatial data structures". *Information Systems* 19, No. 7 (October 1994) : 569-582.

Westherall, B. "Cartography in a changing world", Paper presented at thirteenth annual meeting of the North American Cartographic Information Society, Silver Spring, M.D. October 20-23, 1993). *Geography and Map Division Bulletin* No. 176 (June 1994) : 30-1.

Williamson, Nancy J. "Is there a catalog in your future? Access to information in the year 2006". *Library Resources and Technical Services* 26 (April-June 1982) : 122-35.

Williamson, Nancy J. "Subject Access in the online environment". *Advances in librarianship* 13 (1984) : 47-97.

Wong, M.A. "Exploring the impact of digital cartographic data on map librarianship using data use models". *Geography and Map Division Bulletin* 173 (September 1993) : 2-14.

Wood, A.A. "Simposium or cartographic designs and research and Canadian Cartographic Association/North American Cartographic Information Society", Report of Meetings held in Ottawa Ontario, Canada. August 7-12, 1994. *Geographic Map Division Bulletin,* No. 178 (December 1994) : 57-63.

Woodruff, A.G. and C. Plant. "GIPSY : Automated geographic indexing of text documents". *American Society for Information Science* 45, No. 9 (October 1994) : 645-655.

Yavarkopvsky, 7. and M. Hedstrom. "Development of a prototype state government locator systems". Albany, N.Y. 12230, U.S.A. : New York State Library Cultural Education Centre .

Yott, Patrick. "Mapping 1990 census data". *Information Technology and Libraries* 13, No. 1 (March 1994) . 63-70.

Index